AF581005

# Springer Tracts in Modern Physics
# Volume 213

Starting with Volume 165, Springer Tracts in Modern Physics is part of the [SpringerLink] service. For all customers with standing orders for Springer Tracts in Modern Physics we offer the full text in electronic form via [SpringerLink] free of charge. Please contact your librarian who can receive a password for free access to the full articles by registration at:

springerlink.com

If you do not have a standing order you can nevertheless browse online through the table of contents of the volumes and the abstracts of each article and perform a full text search.

There you will also find more information about the series.

# Springer Tracts in Modern Physics

Springer Tracts in Modern Physics provides comprehensive and critical reviews of topics of current interest in physics. The following fields are emphasized: elementary particle physics, solid-state physics, complex systems, and fundamental astrophysics.
Suitable reviews of other fields can also be accepted. The editors encourage prospective authors to correspond with them in advance of submitting an article. For reviews of topics belonging to the above mentioned fields, they should address the responsible editor, otherwise the managing editor.
See also springeronline.com

Vladimir G. Baryshevsky
Ilya D. Feranchuk
Alexander P. Ulyanenkov

# Parametric X-ray Radiation in Crystals

## Theory, Experiments and Applications

With 63 Figures

Professor Vladimir G. Baryshevsky
Research Institute for Nuclear Problems
Bobruiskaya Str. 11
220050 Minsk, Belarus
E-mail: bar@inp.minsk.by

Dr. Alexander P. Ulyanenkov
Bruker AXS GmbH
Östliche Rheinbrückenstr. 50
76187 Karlsruhe, Germany
E-mail: alex.ulyanenkov@bruker-axs.de

Professor Ilya D. Feranchuk
Department of Physics
Belarusian University
F. Skaryny av.
220050 Minsk, Belarus
E-mail: fer@open.by

Library of Congress Control Number: 2005930452

Physics and Astronomy Classification Scheme (PACS):
41.75.Ht, 41.60.-m, 07.85.Fv, 78.70.-g, 41.60.Bq

ISSN print edition: 0081-3869
ISSN electronic edition: 1615-0430
ISBN-10 3-540-26905-3 Springer Berlin Heidelberg New York
ISBN-13 978-3-540-26905-2 Springer Berlin Heidelberg New York



Springer is a part of Springer Science+Business Media
springeronline.com

Printed in The Netherlands



Typesetting: by the author using a Springer LaTeX macro package
Cover concept: eStudio Calamar Steinen
Cover production: *design &production* GmbH, Heidelberg

Printed on acid-free paper SPIN: 10983438 57/TechBooks 5 4 3 2 1 0

# Preface

The periodic arrangement of atoms (nuclei) influences essentially the electromagnetic processes accompanying the moving charged particles in crystals and results in qualitatively new coherent and orientational effects, which are not observable in amorphous media. Several of these effects were described and systemized in the monograph by M.L. Ter-Mikaelian [1] in 1960s, where the coherent bremsstrahlung, resonant radiation and electron–positron pair creation at high energies were considered. The particles taking part in these processes have a wavelength smaller than the period of crystallographic lattice.

In 1971, a qualitatively different mechanism of electromagnetic radiation from high-energy electrons in the crystal was predicted, when the wavelength of the emitted photons is comparable to the lattice period and the diffraction of the radiation plays a crucial role. The spectrum of this radiation, named parametric X-ray radiation (PXR), depends essentially on the crystallographic parameters. The predicted phenomenon was experimentally observed in 1985, and up to date there has been much theoretical and experimental work done on PXR in numerous scientific centres.

In the present monograph, a systematic description of PXR is given and an analysis of the published studies on PXR is performed. In Part I of the book, the qualitative features of PXR and the difference between PXR and other radiation mechanisms are given along with the methods for PXR simulation under different experimental conditions. In Part II, the experimental results and their theoretical interpretations are discussed (Chaps. 5 and 6). The effective application of PXR phenomenon for modern scientific and technological needs is an actual task of today's investigations, and the prospective applications are considered in Chap. 7.

The authors are indebted to the colleagues at Belarussian State University and Tomsk Polytechnical Institute for a long-term and fruitful collaboration, which resulted in pioneering observations of PXR and a detailed investigation of PXR features.

We thank A.A. Gurinovich for invaluable help in the preparation of the the manuscript, O.M. Lugovskaya for numerical simulations, and Professor V.N. Zabaev and Professor H. Backe for providing the experimental data used in the book.

Finally, we would like to thank International Scientific Technical Centre (under project B-626) and Bruker AXS (Karlsruhe, Germany) for the permanent support of studies on PXR from low-energy electrons.

## References

1. M.L. Ter-Mikaelian: *High Energy Electromagnetic Processes in Condensed Media* (in Russian: AN ArmSSR, Yerevan 1969) (in English: Wiley, New York 1972)

Minsk, Belarus — *Vladimir Baryshevsky*
Minsk, Belarus — *Ilya Feranchuk*
Karlsruhe, Germany — *Alex Ulyanenkov*
April 2005

# Contents

## Part I Theory

# List of Acronyms

| | |
|---|---|
| PXR | parametric X-ray radiation |
| CBS | coherent bremsstrahlung |
| DBS | diffracted bremsstrahlung |
| IBS | incoherent bremsstrahlung |
| HRPXR | high-resolution parametric X-ray radiation |
| LRS | low-resolution scale |
| HRS | high-resolution scale |
| DTR | diffracted transition radiation |
| VCR | Vavilov–Cherenkov radiation |
| FDPXR | forward direction parametric X-ray radiation |
| HRXRD | high-resolution X-ray diffraction |
| PGR | parametric $\gamma$-radiation |
| EAD | extremely asymmetric diffraction |
| CCD | charge couple device |
| FWHM | full width at half maximum |
| PSD | position sensitive detector |
| SR | synchrotron radiation |
| ASM | anomalous scattering method |
| ChR | characteristics radiation |
| FEL | free-electron laser |
| QED | quantum electrodynamics |
| PT | perturbation theory |
| RR | resonant radiation |

# Part I

# Theory

# 1

# Electromagnetic Radiation from a Charged Particle in Crystals: Qualitative Consideration

## 1.1 Optical and X-ray Cherenkov Radiation in Homogeneous Media

Electromagnetic radiation caused by motion of a charged particle has been studied for more than 100 years. It was firmly believed for an appreciable period of time that only accelerated particles can radiate electromagnetic waves.

The possibility of light radiation by a charged particle moving with a constant velocity in a medium began appearing only in 1934 with the works of Vavilov and Cherenkov [12] and Tamm and Frank [167], who explained Cherenkov's experiments. The Vavilov–Cherenkov radiation (VCR) was the first radiation phenomenon in physics that depends not only on the charge and velocity of the particle but also on the optical properties of the medium. Moreover, a fast electron radiating in a medium appeared as the first coherent self-radiating source of light, whose size significantly exceeded the radiation wavelength [99]. Later, numerous confirmations of coherent influence of atoms of a medium on the probability of electromagnetic processes have been discovered and described (see, for example, [21,30]).

A charged particle moving with a constant velocity $\boldsymbol{v}$ in an optically transparent medium with the dielectric permittivity $\varepsilon(\boldsymbol{k},\omega)$, where $\boldsymbol{k}$ is the wave vector of the photon and $\omega$ is its frequency, can radiate if the following condition is fulfilled:

$$1-\frac{v}{c}\,n(\boldsymbol{k},\omega)\cos\theta=0\,, \tag{1.1}$$

where $\theta$ is the angle between $\boldsymbol{v}$ and $\boldsymbol{k}$; $c$ is the light velocity in vacuum and $n(\boldsymbol{k},\omega)=\sqrt{\epsilon(\boldsymbol{k},\omega)}$ is the refractive index of the medium. Evidently, condition (1.1) can only be satisfied if the velocity of the particle is higher than the phase velocity of light in the medium, $c_{\mathrm{m}}$:

$$v>c_{\mathrm{m}}=\frac{c}{n},\qquad n>1\,. \tag{1.2}$$

For a long time, the studies and applications of the VCR were limited within the domain of optical and soft X-ray radiation, when condition (1.2) is satisfied for numerous media and for a wide range of particle energies. Moreover, the VCR channel was assumed to be forbidden in the frequency range higher than characteristic atomic frequencies, where the dielectric permittivity of a uniform medium is determined by the universal expression [25]:

$$\epsilon(\boldsymbol{k},\omega) = 1 - \frac{\omega_0^2}{\omega^2} < 1,\ \omega > \omega_0 . \tag{1.3}$$

Here $\omega_0$ is the plasma frequency of the medium, which is usually within the optical range. Formula (1.3) is not applicable for an anomalous dispersion region, corresponding to internal atomic shells, and for resonant transitions of Mössbauer nuclei. As shown for the first time in [24, 27], in the medium of heavy atoms the VCR is also possible in the range of X-ray wavelengths. However, it is difficult to observe this effect because of the strong radiation absorption in the region of anomalous dispersion.

Analysis of the induced effect of Vavilov–Cherenkov and Doppler effect with transition radiation done in 1971 [4] revealed that conditions for generation of radiation from a charged particle in a crystal (considering that the Bragg diffraction is allowed) fundamentally differ from radiation conditions in an amorphous medium. In this case, the crystal cannot be described by a certain refractive index, because it has several indices of refraction depending on the photon frequency and the direction of the photon propagation. This brings into picture a new radiation type, named parametric X-ray radiation (PXR) [7, 8], from a charged particle moving with a constant velocity in a crystal [4, 6, 19, 30]. Solely the crystal structure of the medium was shown to cause a diffraction of the X-ray bremsstrahlung and transition radiation [4]. Moreover, a relativistic charged particle moving with a constant velocity in a crystal radiates X-rays even at large angles [6], whereas radiation in an amorphous medium is emitted within the angle $\theta \sim 1/\gamma$, where $\gamma = E/mc^2$ is the Lorentz factor of the particle, $E$ is its energy, $m$ is the mass of the particle. The PXR was found out to form both the waves with the refractive index $n > 1$ (slow waves) and the waves with $n < 1$ (fast waves). The X-ray quanta in PXR are emitted at both large and small angles with respect to the particle velocity [6].

The theory of radiation from a charged particle moving with a constant velocity was also considered in [29], where the resonant radiation in a thin crystal was described. In this work, however, not much attention is paid to the refraction effects due to the absence of fast and slow waves and radiation along the particle velocity. Contrary to the PXR case, the frequency of resonant radiation increases with the increase in the particle energy.

Papers [4, 6] initiated numerous publications considering X-ray radiation from a charged particle moving with a constant velocity in a crystal. Despite the numerous theoretical works on PXR, the phenomenon was experimentally observed for the first time in 1985 [1,5]. The results of the experiments were in good quantitative agreement with theoretical predictions. Presently, the PXR physics and applications are being studied in numerous scientific centres and essential volume of accumulated information requires a systematization and general analysis, which is the main goal of the present monograph.

## 1.2 Pseudophoton Spectrum of a Relativistic Electron

The most pictorial qualitative description of the PXR mechanism is given by the electrodynamic pseudophoton concept introduced in [32]. This approach exploits the approximate equivalence of the energy flux of the electromagnetic field, created by a constantly moving charged particle on the one hand, and the energy flux transferred by the photon beam with a certain angular and spectral distribution, on the other hand. To calculate this spectrum, the solution for Maxwell's equation has to be found for the vector $\boldsymbol{A}(\boldsymbol{r},t)$ and scalar $\varphi(\boldsymbol{r},t)$ potentials, determined from a classic expression for the current $\boldsymbol{j}(\boldsymbol{r},t)$ and charge $\rho(\boldsymbol{r},t)$ density of a point particle with the charge $q = Ze$ [2]:

$$\begin{aligned} \left(\Delta - \frac{1}{c^2}\frac{\partial^2}{\partial t^2}\boldsymbol{A}(\boldsymbol{r},t)\right) &= -\frac{4\pi}{c}\boldsymbol{j}(\boldsymbol{r},t)\ , \\ \left(\Delta - \frac{1}{c^2}\frac{\partial^2}{\partial t^2}\varphi(\boldsymbol{r},t)\right) &= -4\pi\rho(\boldsymbol{r},t)\ , \\ \boldsymbol{j}(\boldsymbol{r},t) = \boldsymbol{v}\rho(\boldsymbol{r},t),\ \ \rho(\boldsymbol{r},t) &= q\delta(\boldsymbol{r}-\boldsymbol{v}t)\ . \end{aligned} \tag{1.4}$$

Using a standard expansion of potentials and the $\delta$-function in Fourier integrals, the following expressions for electric and magnetic fields accompanying the particle are derived (the $z$-axis is along the velocity vector $\boldsymbol{v}$):

$$\begin{aligned} \boldsymbol{E}(\boldsymbol{r},t) = -\frac{1}{c}\frac{\partial \boldsymbol{A}}{\partial t} - \nabla\varphi &= -\mathrm{i}\frac{q}{2\pi^2}\int \mathrm{d}\boldsymbol{k}\frac{\boldsymbol{k}c^2 - \boldsymbol{v}(\boldsymbol{k}\boldsymbol{v})}{k^2c^2 - k_z^2v^2}\,\mathrm{e}^{\mathrm{i}\boldsymbol{k}\boldsymbol{r}-\mathrm{i}k_z vt}\ , \\ \boldsymbol{H}(\boldsymbol{r},t) &= \frac{1}{c}[\boldsymbol{v}\boldsymbol{E}(\boldsymbol{r},t)]\ . \end{aligned} \tag{1.5}$$

In the case of an ultrarelativistic particle with the velocity $v \approx c$, the parameter $\omega = k_z v$, defining the time dependence in (1.5), can be considered as a frequency of a pseudophoton, and the electromagnetic field in (1.5) can be split into longitudinal $E_z$ (along $\boldsymbol{v}$) and transverse $E_\perp$ components:

$$\begin{aligned} \boldsymbol{E}_\perp &= -\mathrm{i}\frac{q}{2\pi^2 v}\int \mathrm{d}\boldsymbol{k}_\perp \mathrm{d}\omega\frac{\boldsymbol{k}_\perp}{k_\perp^2 + \omega^2(1/v^2 - 1/c^2)}\mathrm{e}^{\mathrm{i}\boldsymbol{k}_\perp\boldsymbol{r}_\perp + \mathrm{i}\omega(z/v-t)}\ , \\ \boldsymbol{E}_z &= -\mathrm{i}\frac{q}{2\pi^2 v^2}\int \mathrm{d}\boldsymbol{k}_\perp \mathrm{d}\omega\frac{\omega(1-v^2/c^2)}{k_\perp^2 + \omega^2(1/v^2 - 1/c^2)}\mathrm{e}^{\mathrm{i}\boldsymbol{k}_\perp\boldsymbol{r}_\perp + \mathrm{i}\omega(z/v-t)}\ , \\ \boldsymbol{H}(\boldsymbol{r},t) &= \frac{1}{c}[\boldsymbol{v}\boldsymbol{E}(\boldsymbol{r},t)]\ . \end{aligned} \tag{1.6}$$

We used here a standard definition of the Lorentz factor for the particle of energy $E$ and mass $m$:

$$\gamma = \frac{E}{mc^2} = \frac{1}{\sqrt{1 - v^2/c^2}} . \tag{1.7}$$

For an ultrarelativistic particle ($\gamma \gg 1$), the transverse component of the electromagnetic field is essentially larger than the longitudinal one $|\boldsymbol{E}_\perp| \gg E_z$. This situation approximately corresponds to the superposition of free electromagnetic waves with frequencies $\omega$ propagating along the vector $\boldsymbol{v}$. The energy flux $\Pi_z$ transferring the electromagnetic field of the particle (1.6) into the direction of its velocity is

$$\begin{aligned}\Pi_z &= \int \mathrm{d}\boldsymbol{r}_\perp \int_{-\infty}^{\infty} \mathrm{d}t \frac{c}{4\pi} [\boldsymbol{E}\boldsymbol{H}]_z = \int \mathrm{d}\boldsymbol{r}_\perp \int_{-\infty}^{\infty} \mathrm{d}t \frac{v}{4\pi} |\boldsymbol{E}_\perp|^2 \\ &= \frac{q^2}{4\pi^2 v} \int \mathrm{d}\boldsymbol{k}_\perp \mathrm{d}\omega \frac{k_\perp^2 \,\mathrm{d}\omega}{[k_\perp^2 + \omega^2/\gamma^2 v^2]^2} . \end{aligned} \tag{1.8}$$

However, the energy flux can be represented as the flux of pseudophotons with energy $\hbar\omega$ and angular–spectral distribution $n(\theta, \omega)$, which is integrated over frequencies and angles:

$$\Pi_z = \int \mathrm{d}\boldsymbol{k}_\perp \mathrm{d}\omega \hbar\omega n(\theta, \omega) , \tag{1.9}$$

where the angle $\theta$ between the wave vector of the pseudophoton and the particle velocity $\boldsymbol{v}$ is established by the relation $\sin\theta = k_\perp/\omega$. Comparing (1.8) and (1.9), the spectral–angular distribution of pseudophotons is found to be

$$n(\theta, \omega) = \frac{q^2}{4\hbar\omega\pi^2 v} \frac{\sin^2\theta}{[\sin^2\theta + 1/\gamma^2]^2} . \tag{1.10}$$

The well-known expression for the frequency distribution of pseudophotons is obtained from (1.10) if the integration over $k_\perp$ is performed [2]:

$$n(\omega) = \frac{q^2}{2\hbar\omega\pi v} \ln\left(\frac{\eta\gamma mc^2}{\hbar\omega}\right) . \tag{1.11}$$

Here, the parameter $\eta \sim 1$ influences inessentially the spectrum and is determined from the applicability condition for the point particle model [2].

Figure 1.1 shows the angular distribution of pseudophotons for several different values of the energy of the charged particle, $n_0 = \frac{q^2}{4\pi^2 v}$. The essential feature of the functions shown is that they are independent of the frequency. Each distribution corresponds to the pseudophoton beam, concentrated within the cone $\Delta\theta \approx \gamma^{-1}$ with typical minimum along the vector $\boldsymbol{v}$. Thus, for $\gamma \gg 1$ each pseudophoton can be assumed to have a certain wave vector with a high accuracy:

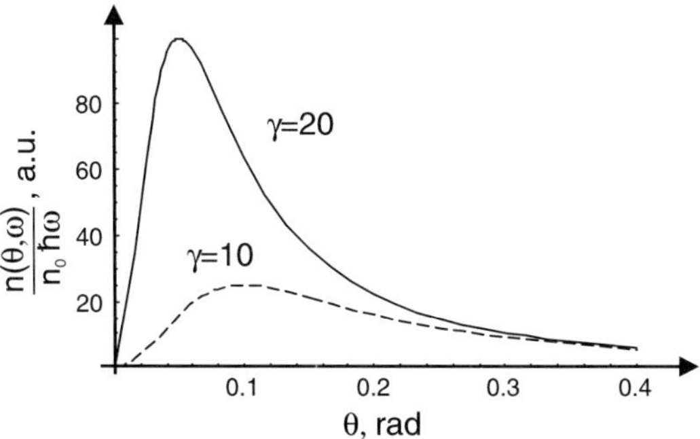


**Fig. 1.1.** Characteristic angular distribution of pseudophotons

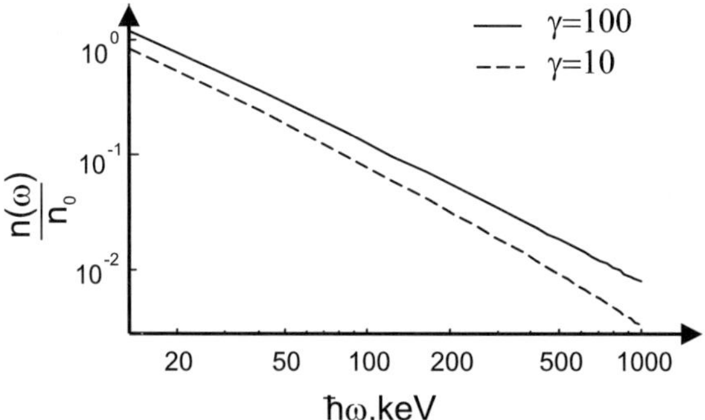


**Fig. 1.2.** Characteristic spectral distribution of pseudophotons

$$\boldsymbol{k} = \omega \frac{\boldsymbol{v}}{v^2} \, . \tag{1.12}$$

At the same time, as Fig. 1.2 shows, the frequency spectrum with the weight $\sim \omega^{-1}$ contains pseudophotons of arbitrary energy, not exceeding the energy of the particle. Thus, the electromagnetic field of the relativistic charged particle moving in vacuum is equivalent to that of the narrow beam of white electromagnetic radiation with a broad frequency spectrum.

For the particle moving in a medium, Maxwell's equations (1.4) have to be corrected for polarizability of the material and for elastic and inelastic collisions of the particle and atoms, which influence the electromagnetic field of the particle. However, for high-energy particles and thin samples, these corrections are negligible and can be considered in the framework of perturbation theory. Then the emission process from the charged particle in a medium is equivalent to the scattering of pseudophotons by atoms. The calculation of the flux of scattered photons gives a differential cross-section of radiation, which is a key idea of the pseudophoton approach.

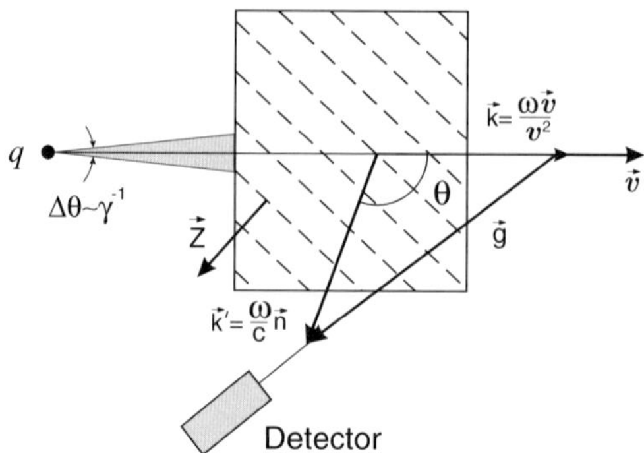


**Fig. 1.3.** Radiation process as the coherent scattering of pseudophotons

## 1.3 Coherent Bremsstrahlung, Resonant Radiation and Parametric Radiation in Crystals

Coherent bremsstrahlung (CBS) is one of the famous orientational radiation effects accompanying a relativistic particle moving in a crystal. This radiation was shown for the first time in [14, 18, 29] to be caused by interference of the photons emitted by a charged particle, which interacts with periodically arranged atoms of the crystal. The phenomenon of CBS is being well studied nowadays, and there is a diversity of theoretical and experimental works, the results of which are systematized in the monograph [30]. Another radiation mechanism caused by the periodical properties of the medium is resonant radiation (RR). It appears as a result of the interference of the transition radiation on the boundary of a one-dimensional periodic medium (see, for example, [15,26,30]) and is similar to the Smith–Purcell radiation [28]. PXR, being also the orientational radiation effect, differs essentially from both CBS and RR.

The pseudophoton approach is used below to qualitatively describe the characteristics and differences of these emission types.

Let us consider a charged particle incident on a crystal as a beam of pseudophotons (Fig. 1.3). The scattering amplitudes of the pseudophotons reflected from the periodical arrangement of atoms in the crystal are coherent in certain directions and create the diffraction peaks, analogous to those produced by scattering of the external electromagnetic field from the crystal. These diffraction peaks [23] can be detected if the wave vector $\boldsymbol{k}'$ of the scattered radiation is connected with the wave vector $\boldsymbol{k}$ of the incident beam by the following formula:

$$\boldsymbol{k}' = \boldsymbol{k} + \boldsymbol{g}; \qquad \frac{\omega}{c}\boldsymbol{n} = \frac{\omega \boldsymbol{v}}{v^2} + \boldsymbol{g} \ . \tag{1.13}$$

This relationship can satisfy any vector $\boldsymbol{g}$ from the set of the reciprocal lattice vectors. Vectors $\boldsymbol{g}$ are defined by a translational symmetry of the crystal and are an important characteristic of any diffraction process which occurs in the crystal [23].

Formula (1.13) takes into account the conversion of the primary wave pseudophoton with the wave vector $\boldsymbol{k}$ in (1.12) into a real photon $\boldsymbol{k}'$ with the frequency $\omega$, emitted in the direction parallel to the unit vector $\boldsymbol{n}$. The electron velocity does not change during this process (Fig. 1.3). As follows from (1.13), the coherent process is impossible in a homogeneous medium when $\boldsymbol{g} = 0$. In contrast to the diffraction of external radiation, when the frequency and velocity of a primary photon are clearly defined (1.13) is considered as the equation for the spectrum of the emitted photons. For a certain value of the particle velocity and $\boldsymbol{g} \neq 0$, the frequency spectrum of the photons follows from the solution of the quadratic equation resulting from (1.13):

$$\frac{\omega^2}{v^2\gamma^2} + 2\frac{\omega}{v^2}(\boldsymbol{vg}) + g^2 = 0 . \qquad (1.14)$$

As a result, two branches of the spectrum of coherent radiation from a charged particle in the crystal are distinguished:

$$\omega_{1,2} = \gamma^2 \left\{ -(\boldsymbol{vg}) \pm \sqrt{(\boldsymbol{vg})^2 - \frac{v^2 g^2}{\gamma^2}} \right\} . \qquad (1.15)$$

For high energies ($\gamma \gg 1$), this equation is transformed to

$$\begin{aligned} &\omega_1 \approx 2|\boldsymbol{vg}|\gamma^2 \approx \frac{4\pi}{l}\gamma^2 c ; \\ &\omega_2 \approx \frac{v^2 g^2}{2|(\boldsymbol{vg})|} = \frac{\pi g v}{\sin\theta_{\mathrm{B}}} ; \qquad \sin\theta_{\mathrm{B}} = \frac{|(\boldsymbol{vg})|}{vg} . \end{aligned} \qquad (1.16)$$

Thus, the wavelength of photons in the first branch ($\lambda_1 = 2\pi c/\omega$) is essentially (with a factor $\sim \gamma^{-2}$) less than the lattice period, whereas the wavelength of the second branch is of the same order of magnitude as the distance between the crystallographic planes corresponding to the vector $\boldsymbol{g}$.

In each branch, the photons are emitted within the narrow cones, axes of which are differently directed (Fig. 1.4a):

$$\begin{aligned} \cos\theta_1 &= \frac{(\boldsymbol{vn})}{v} = \frac{c}{v} + \frac{(\boldsymbol{vg})}{v\omega_1} \approx 1 ; \\ \cos\theta_2 &= \frac{c}{v} + \frac{(\boldsymbol{vg})}{v\omega_2} \approx 1 - 2\sin^2\theta_{\mathrm{B}} ; \\ \theta_2 &\approx 2\theta_{\mathrm{B}} . \end{aligned} \qquad (1.17)$$

The first branch corresponds to RR, and for relativistic particles the photons are emitted at small angles to the vector $\boldsymbol{v}$. The wavelength of the photons is essentially less than the lattice period and decreases with increasing particle energy $\sim \gamma^2$. The second branch is the PXR, which is emitted into directions

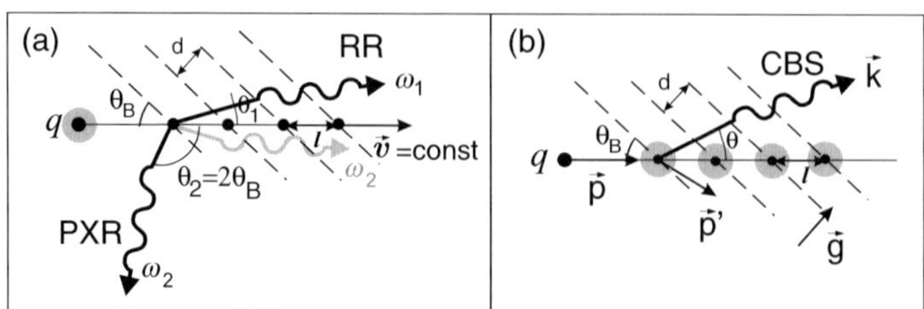


**Fig. 1.4.** (**a**) Formation of RR (interference of the pseudophotons from the *charged particle* field under small angle scattering) and PXR (interference of pseudophotons under specular reflection from the planes; the forward PXR is shown in Sect. 6.2); (**b**) formation of CBS (scattering of the pseudophotons from the *atomic field* by the charged particle)

making a large angles to the vector $\boldsymbol{v}$ and the frequency of radiation is determined by the crystallographic unit cell and almost independent of the electron energy.

In terms of X-ray diffraction [3], the physics of RR corresponds to the small-angle diffraction of the hard region of the pseudophoton spectrum, whereas PXR is related to the Bragg diffraction of pseudophotons with the frequency of X-ray radiation. Thus, the angular distribution of PXR is represented by a set of peaks (reflections), corresponding to the diffraction of X-ray beam in the crystal; this beam has an angular dispersion $\gamma^{-1}$ and frequency spectrum (1.11). Each reflection, corresponding to a certain crystallographic plane, defined by the reciprocal lattice vector $\boldsymbol{g}$, is the X-ray beam with the angular width $\gamma^{-1}$ travelling at an angle $\theta_B$ to the particle velocity $\boldsymbol{v}$:

$$\theta_{\mathrm{B}} = \arcsin\left(\frac{|(\boldsymbol{vg})|}{vg}\right) . \tag{1.18}$$

The emission of photons at large angles to the velocity of a relativistic particle makes a principal difference between PXR and other radiation types (incoherent bremsstrahlung, channelling radiation, undulator and transition radiation, etc.), which are localized in the narrow cone of width $\gamma^{-1}$ around velocity $\boldsymbol{v}$. PXR reflections have a spectral width $\Delta\omega/\omega \approx \gamma^{-1}$ and are located near the frequencies

$$\omega_n = \frac{\pi c}{d\sin\theta_{\mathrm{B}}} n, \qquad n = 1, 2\ldots; \qquad \boldsymbol{g} = \frac{2\pi n}{d}\boldsymbol{Z}\,, \tag{1.19}$$

where the unit vector $\boldsymbol{Z}$ is perpendicular to the crystallographic planes (Fig. 1.3). In experimentally measured units, (1.19) is written as

$$\hbar\omega_n\ [\mathrm{keV}] = 1.974\frac{\pi}{\mathrm{d}[\mathrm{\AA}]\sin\theta_{\mathrm{B}}} n\,. \tag{1.20}$$

We consider here the fundamental distinctions between the formation mechanisms of PXR and CBS. As mentioned in [2, 30], bremsstrahlung is emitted due to scattering of the pseudophotons from the atom electromagnetic field by the moving charged particle (Fig. 1.4b). PXR, in contrast, arises when the pseudophotons from the electromagnetic field produced by the charged particle are scattered by the atom electrons. Bremsstrahlung is always accompanied by a change in the momentum of the incident particle and the bremsstrahlung spectrum is determined by the conservation law:

$$\begin{aligned} \boldsymbol{p} - \boldsymbol{p}' - \hbar \boldsymbol{k} &= \boldsymbol{u} + \boldsymbol{g} \, , \\ E - E' - \hbar\omega &= 0 \, . \end{aligned} \tag{1.21}$$

When $\boldsymbol{g} = 0$, each of the atoms of the medium acquires the minimal transferred momentum $u$, which corresponds to the conventional incoherent Bethe–Heitler bremsstrahlung [30]. When $\boldsymbol{g} \neq 0$ (the vector $\boldsymbol{u}$ can be zero), the momentum is transferred to the crystal as a whole and condition (1.21) defines the CBS spectrum [30]. The system of equations (1.21) has a single solution

$$\omega_n^{(\mathrm{CBS})}(\theta) = \frac{\boldsymbol{g}\boldsymbol{v}}{1 - v/c\cos\theta} = \frac{2\pi n}{d}\,\frac{v\sin\theta_{\mathrm{B}}}{1 - v/c\cos\theta} \, . \tag{1.22}$$

Therefore, kinematically the CBS can be emitted at an arbitrary $\theta$. However, the particle energy is high ($\gamma \gg 1$) and the radiation intensity increases for the small angles $\theta \sim \gamma^{-1}$ [30], thus resulting in $\omega_n^{(\mathrm{CBS})} \sim \gamma^2 \frac{c}{d}$; similar results are found for RR. For $\gamma \sim 1$, CBS can exist at large angles and interfere with PXR (see Chap. 4).

The detailed analysis of PXR spectra is performed in the following sections. In this section, we consider the restrictions which result from the approximations used for the description of this phenomenon by means of the pseudophoton concept. The first approximation used assumes neglecting quantum recoil during the photon emission. This approximation results from Maxwell's equations for a point particle moving along the classical trajectory, and the assumption is valid when the energy of the photon is much less than the particle energy:

$$\hbar\omega \ll E \, . \tag{1.23}$$

The second approximation is the use of the kinematic (perturbation) theory for pseudophoton diffraction. As for conventional X-ray diffraction, this approximation is valid for thin crystals of thickness $L$, which is less than the extinction length $L_{\mathrm{ext}}$, defined by the frequency of radiation and crystal polarizability [3]:

$$L < L_{\mathrm{ext}} = \frac{c}{\omega|\epsilon - 1|} \, . \tag{1.24}$$

In the X-ray domain, the polarizability $|\epsilon - 1| \sim 10^{-4}$–$10^{-5}$; thus, the kinematic theory of PXR is valid in the crystals with the thickness $L \sim 1$–$10\,\mu\mathrm{m}$.

The limitation for the crystal thickness is also caused by multiple (incoherent) scattering of the charged particle in a crystal. For relativistic particles, the thickness limit can be estimated from the root mean square of the angle $\theta_s^2(L)$, which defines the deviation of the particle velocity from the primary direction [30]:

$$\theta_s^2(L) = \frac{E_s^2}{E^2}\frac{L}{L_R}\,, \tag{1.25}$$

where $E_s \approx 21\,\mathrm{MeV}$ and $L_R$ is the radiation length.

As follows from (1.10), the typical angular width of the pseudophoton beam is defined by the parameter $\gamma^{-1}$. Therefore, the deviation of the particle trajectory due to multiple scattering does not influence the pseudophoton spectrum and, consequently, the spectrum of the emitted photons if the following conditions are fulfilled:

$$\theta_s^2(L) < \gamma^{-2}; \qquad L < \left(\frac{mc^2}{E_s}\right)^2 L_R \approx 6.3 \times 10^{-4} L_R\,. \tag{1.26}$$

In most of the crystals, these conditions are easier than limitation (1.24), i.e. dynamical diffraction effects show up for the crystal thickness for which the multiple scattering is still negligible. In a similar way, the influence of mosaicity on the PXR spectrum can be evaluated: The parameter describing mosaicity has to be smaller than the angular distribution of the pseudophotons $\sim\gamma^{-1}$.

## 1.4 Pioneering Experiments on the Observation of PXR

After the theoretical prediction of PXR in 1971, the detailed quantitative description of PXR characteristics was given in a series of posterior publications [8–11, 16, 17, 20]. However, the first successful experiments on the observation of PXR were carried out 14 years later in 1985 by the joint experimental team from Tomsk and Minsk. In the first experiment, the spectrum of the photons emitted within PXR high angular reflection was measured [1,13]. The experiment was carried out at the internal beam of the synchrotron facility Sirius in Tomsk, Russia. The geometry of the experiment is shown in Fig. 1.5. The electron beam with the angular divergence of 0.1 mrad and monochromaticity of 0.5% was incident on a diamond target with the dimensions $10 \times 6 \times 0.35\,\mathrm{mm}^3$, which was fixed in a two-axis goniometer. The beam pulse duration was $\tau_0 = 15$ ms, and the energy of the electron beam was $E = 900\,\mathrm{MeV}$.

Since the angular divergence of the studied X-ray radiation should be of the order $\gamma^{-1}$ (in the considered experiment it was 0.6 mrad), the matching of a Bragg reflection from any crystallographic plane with a detector axis is a

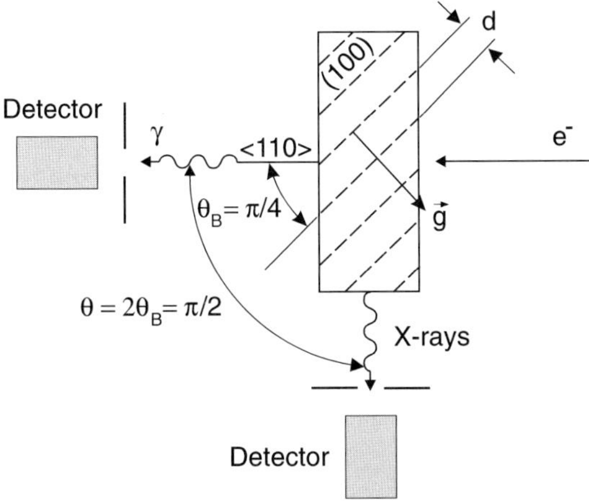


**Fig. 1.5.** Geometry of the first PXR experiment

difficult problem. In the experiment, the detector aperture varied from 6 to 20 mrad, that is more than $\gamma^{-1}$. To resolve this problem, the diamond crystal was cut off perpendicularly to the $\langle 100 \rangle$ axis, and the detector was placed at the angle $\pi/2 \pm 3 \times 10^{-3}$ rad relative to the electron beam. In this case, the $\langle 100 \rangle$ crystallographic axis is aligned to incident electron momentum for the electron beam intersecting the (100) planes at the angle $\theta_{\mathrm{B}} = \pi/4$ and, therefore, the monochromatic X-ray radiation is expected at the angle $2\theta_{\mathrm{B}} = \pi/2$, at which the detector was set up. The crystallographic axis of the target was aligned with the electron beam using the channelling radiation.

To measure the spectral and orientational characteristics of the X-ray radiation, both a NaI(Tl) scintillation spectrometer with a crystal thickness of about 1 mm and a proportional counter filled with xenon were used. The 50-mm diameter entrance window was made of 300-μm-thick beryllium foil. The energy resolution of the scintillation spectrometer at the $^{57}$Co line ($\hbar\omega_\gamma = 14.4$ keV) was $\Delta\omega/\omega \approx 35\%$, and that of the proportional counter was $\Delta\omega/\omega \approx 12\%$. The energy threshold was about $\hbar\omega_{\mathrm{th}} \approx 12$ keV for the NaI(Tl) detector and $\hbar\omega_{\mathrm{th}} \approx 3$ keV for the proportional counter.

The X-ray radiation spectrum measured by the scintillation spectrometer is shown in Fig. 1.6, where the peak at $\hbar\omega_{\mathrm{exper}} = 19.5 \pm 0.3$ keV is clearly seen. The theoretical value of the expected PXR peak can be calculated using (1.20) with the lattice constant $d = 3.57\,\text{Å}$ and $\theta_{\mathrm{B}} = \pi/4$, which results in $\hbar\omega_{\mathrm{theor}} = 19.7\,\mathrm{keV}$ for the eighth-order reflection ($n = 8$) from the (100) diamond planes and is in good agreement with the experiment.

The peak width at half maximum is about $\Delta\omega/\omega \approx 30\%$, which agrees with the intrinsic energy resolution of the detector. The measured photon yield at the energy $\hbar\omega = 19.5$ keV (the area under the peak) is about $N_{\mathrm{ph}} \approx 10^{-8}$ photons/electron.

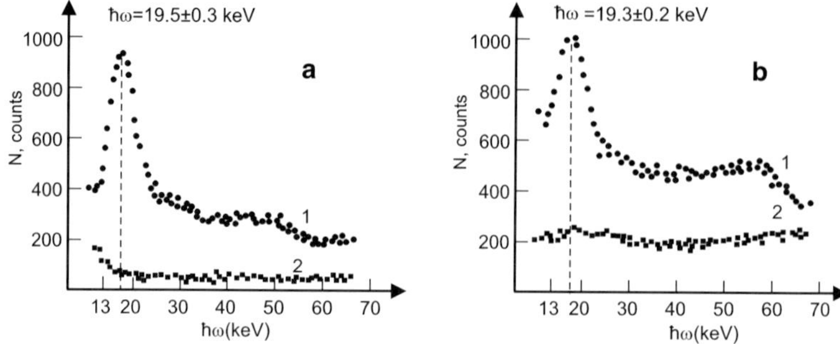


**Fig. 1.6.** The X-ray radiation spectrum of (**a**) 900 MeV and (**b**) 600 MeV electrons from a diamond crystal, measured by the NaI(Tl) spectrometer

To check the correctness of observations, the photon spectrum at the same geometry and for the electron beam energy $E = 600\,\mathrm{MeV}$ was measured. The peak position was not changed (Fig. 1.6b), which confirms the analysis above: The energy of PXR photons is determined by the type and orientation of a crystal target. Another check was done by moving the detector from the 90° Bragg angle to $\theta = 85°$, where no anomalies were observed (the lower curve in Fig. 1.6a). Similar results were obtained by rotating the diamond crystal by 25 mrad keeping the detector position fixed (the lower curve in Fig. 1.6b).

The high-energy threshold of the NaI(Tl) spectrometer did not permit us to observe the lower-order Bragg reflections in the measured spectra. To demonstrate a set of diffraction peaks, the X-ray spectrum was also measured using the proportional counter. Figure 1.7 shows the X-ray radiation spectrum measured in the Bragg geometry for electrons with energy $E = 900\,\mathrm{MeV}$ (the background is eliminated). Two maxima at photon energies $\hbar\omega_{\mathrm{exper}} =$ 19.7 keV and 9.9 keV fitting well the theoretical values were clearly observed.

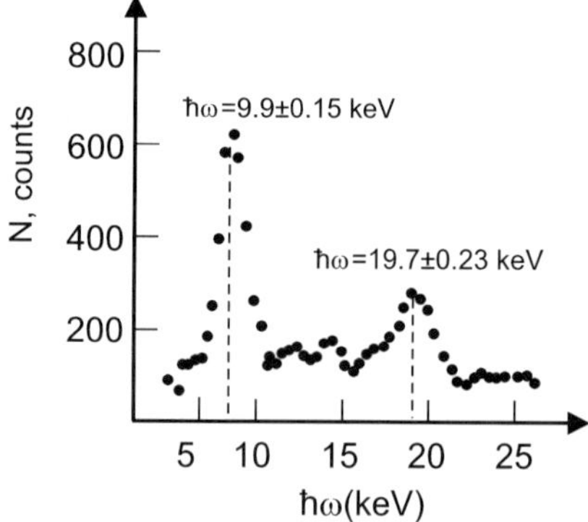


**Fig. 1.7.** The X-ray radiation spectrum of 900 MeV electrons from a diamond crystal, measured by the proportional counter

The width of the peaks made a good agreement with the energy resolution of the proportional counter.

These results gave evidence of the first observation of the PXR in the experiments [1, 13]. It should also be mentioned that the peak at the photon energy $\hbar\omega_{\text{exper}} = 19.5$ keV in the X-ray spectrum from 900 MeV electrons was observed slightly earlier in 1985 [31]. However, that was a single peak emerging just slightly from the background in the emission spectrum. Therefore, this observation was not a reliable basis for drawing the conclusion about PXR, since such a tiny peak might be caused by the background or other side processes.

The important step in PXR studies was reported in the experiment in [5], where the fine angular structure of photon distribution in PXR reflection was measured. This experiment confirmed an adequacy of PXR representation as a diffraction of the pseudophoton beam from the crystallographic planes. According to (1.10) and Fig. 1.1, this beam has distinctive angular distribution with the intensity minimum along the particle velocity and maximum at an angle $\theta_{\text{m}} = \gamma^{-1}$ to the velocity $\boldsymbol{v}$. The theoretical work [17] demonstrated that the reflection of the pseudophoton beam from the crystallographic planes causes the same fine structure in PXR reflection, which was experimentally observed for the first time in [5].

An experimental measurement of the angular distribution of PXR from the 900 MeV electron beam of the synchrotron Sirius (Tomsk, Russia) was reported in [5]. The beam was aligned along the $\langle 100 \rangle$ axis of the diamond crystal with the thickness $L = 0.08$ cm. The X-ray detector was set at the angle $2\theta_{\text{B}} = 90°$ relative to the electron velocity in the plane of the vectors $\boldsymbol{v}$ and $\boldsymbol{g}$, where the reciprocal lattice vector $\boldsymbol{g}$ corresponds to the crystallographic planes (220). Under these conditions, the frequency spectrum of the photons recorded by the detector consists of the set of lines with $\omega^{(n)} = n\omega_{(220)}$; $\hbar\omega_{(220)} = 6.96$ keV.

The measurement of the PXR fine angular distribution was carried out using the coordinate detector made of square cells with linear dimension 1.3 cm. Each of these cells was an ionization camera designed for maximum efficiency of registration for the photons with the energy from 2 to 10 keV. The detector contained $16 \times 16$ cells and was placed at a distance $L_1 = 1$ m from the crystal. The signals from the cells were collected by the computer with a time interval $t$. After the first measurement, the detector was moved two cells along the $x$-axis and the signal was collected with the same interval. The results of both measurements were subtracted from one another in order to decrease the influence of the noise signals. The experimental results for the central cells are shown in Fig. 1.9 and the corresponding theoretical simulations in Fig. 1.8 (the detailed formulas are given in Sect. 2.1). The asymmetry of the PXR angular distribution in both figures confirms the validity of PXR theory. Furthermore, the rotation of the crystal by an angle $\Delta\theta$ in the experiment led to the shift of the intensity maximum for the distance $2L_1\Delta\theta$, which is in agreement with (1.18).

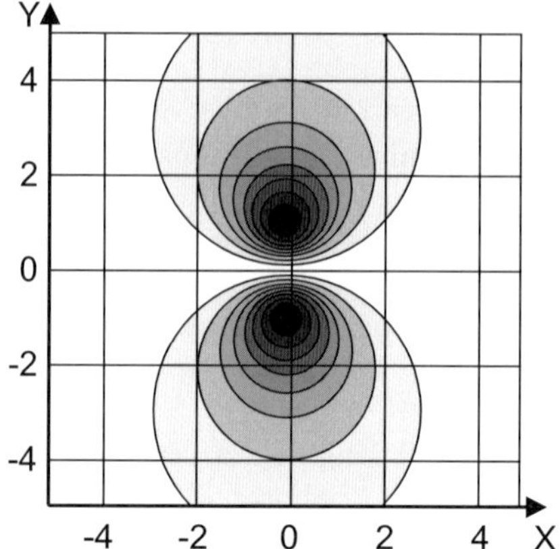


**Fig. 1.8.** Theoretical angular distribution of the photons for a single PXR reflection

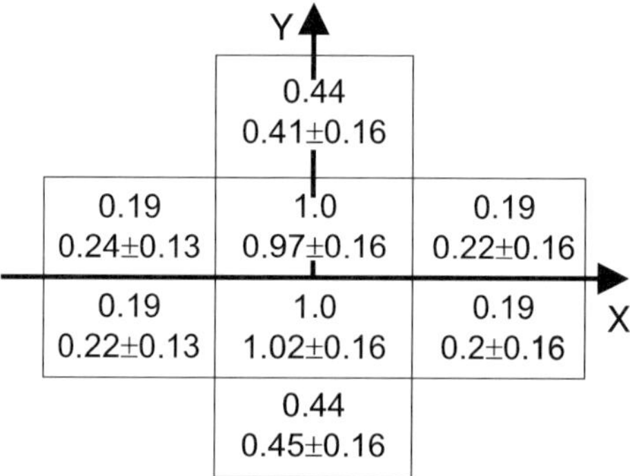


**Fig. 1.9.** Measured (*upper*) and calculated (*lower*) values for the angular distribution of the photons for a single PXR reflection

The average number of photons detected from one electron was also estimated from measurements. Figure 1.9 shows the theoretical (the upper numbers) and experimental (lower numbers) values of photon number from an electron counted on the different cells of the detector. These values were divided by the value $N_{\mathrm{m}}$, which defines the number of photons counted in the two central cells. The experimental value of $N_{\mathrm{m}}$ is $(1.0 \pm 0.2) \times 10^{-6}$ quanta/electron; the theoretical value was estimated as $7.8 \times 10^{-7}$ quanta/electron. Thus, the first experiments were qualitatively and quantitatively well fitted by the PXR theory.

## References

1. Y.N. Adishchev, V.G. Baryshevsky, S.A. Vorobiev, V.A. Danilov, S.D. Pak, A.P. Potylizyn, P.F. Safronov, I.D. Feranchuk: Sov. Phys. JETP Lett. **41**, 361 (1985)

2. A.I. Akhiezer, V.B. Berestetzkii: *Quantum Electrodynamics* (Nauka, Moscow 1969) pp 463–467
3. A. Authier: *Dynamical Theory of X-ray Diffraction* (Oxford University Press, New York 2001) pp 3–26
4. V.G. Baryshevsky: Dokl. Akad. Sci. Belarus **15**, 306 (1971); V.G. Baryshevsky, I.D. Feranchuk: *Proc. XXI Conf. on Nuclear Structure and Nuclear Spectroscopy* (Moscow University, Moscow 1971) p 220
5. V.G. Baryshevsky, V.A. Danilov, O.L. Ermakovich, I.D. Feranchuk, A.V. Ivashin, V.I. Kozus, S.G. Vinogradov: Phys. Lett A **110**, 477 (1985)
6. V.G. Baryshevsky, I.D. Feranchuk: Zh. Eksp. Teor. Fiz. **61**, 944 (1971) (Sov. Phys. JETP **34**, 502 (1972)); Erratum: ibid **64**, 760 (1973) (Sov. Phys. JETP **36**, 399 (1973))
7. V.G. Baryshevsky, I.D. Feranchuk: Izv. Belarusian Acad. Sci. (Ser. Fiz.-Mat. Nauk) **N 2**, 117 (1975)
8. V.G. Baryshevsky, I.D. Feranchuk: Phys. Lett. A **57**, 183 (1976)
9. V.G. Baryshevsky, I.D. Feranchuk: Phys. Lett. A **76**, 452 (1980)
10. V.G. Baryshevsky, I.D. Feranchuk: J. Phys. (Paris) **44**, 913 (1983)
11. V.G. Baryshevsky, I.D. Feranchuk: Nucl. Instrum. Methods **228**, 490 (1985)
12. P.A. Cherenkov: Dokl. Akad. Sci. USSR
13. A.N. Didenko, B.N. Kalinin, S. Pak, A.P. Potylitsin, S.A. Vorobiev, V.G. Baryshevsky, V.A. Danilov, I.D. Feranchuk: Phys. Lett. A **110**, 177 (1985)
14. F. Dyson, H. Uberall: Phys. Rev. **99**, 604 (1955)
15. Ya.B. Fainberg, N.A. Khizhnyak: Zh. Eksp. Teor. Fiz. **32**, 883 (1957) (Sov. Phys. JETP **5**, 720 (1957))
16. I.D. Feranchuk: Kristallografia **24**, 289
17. I.D. Feranchuk, A.V. Ivashin: J. Phys. (Paris) **46**, 1981 (1985)
18. B. Feretti: Nuovo Chimento **7**, 118 (1950)
19. G.M. Garybyan, C. Yang: Zh. Eksp. Teor. Fiz. **61**, 930 (1971) (Sov. Phys. JETP **34**, 495 (1972))
20. G.M. Garibyan, C. Yang: *X-ray Transition Radiation* (Armenian Academy of Sciencies, Erevan 1983)
21. V.L. Ginzburg: *Theoretical Physics and Astrophysics* (Nauka, Moscow 1975) pp 120–160
22. J. Kirz, D.T. Attwood, B.L. Henke: *X-ray Data Booklet* (Lawrence Berkeley Laboratory, Berkeley 2001)
23. Ch. Kittel: *Introduction to Solid State Physics* (Wiley, New York 1977) pp 77–84
24. A. Kolpakov: Yad. Fiz. **16**, 1003 (1972)
25. L.D. Landau, E.M. Lifshitz: *Electrodynamics of Continuous Media* (Nauka, Moscow 1982)
26. V.E. Pafomov: Zh. Tekhn. Fiz **33**, 557 (1963) (Sov. Phys. Tech. Phys. **8**, 412 (1963)); Proc. Lebedev Inst. Phys. **44**, 25 (1971)
27. E.A. Perelshtein, M.I. Podgoretzky: Yad. Fiz. **12**, 1149 (1970)
28. S.L. Smith, E.M. Purcell: Phys. Rev. **91**, 1069 (1953)
29. M.L. Ter-Mikaelian: Zh. Eksp. Teor. Fiz. **25**, 289 (1953)
30. M.L. Ter-Mikaelian: *High Energy Electromagnetic Processes in Condensed Media* (in Russian: AN ArmSSR, Yerevan 1969) (in English: Wiley, New York 1972)
31. A. Vorob'ev, B.N. Kalinin, S. Pak, A.P. Potylitsin: Sov. Phys. JETP Lett. **41**, 1 (1985)
32. C. Weizsäcker: Zs. f. Phys. **88**, 812 (1934); E. Williams: Phys. Rev. **45**, 729 (1934)

# 2

# Radiation from a Charged Particle in Periodic Media: Classical Theory

## 2.1 Representation of Radiation Field via Solution of the Homogeneous Maxwell's Equations

The method of pseudophotons, considered in the previous chapter, is convenient for qualitative analysis of parametric X-ray radiation (PXR). PXR in real crystals, however, has to be quantitatively described more precisely than the pseudophoton concept. In this chapter, the details and the peculiarities of the methods used for simulation of PXR spectra are discussed.

In classic electrodynamics, an electromagnetic radiation is described on the basis of the constrained system of Maxwell's equations for the electromagnetic field and the motion equations for the particle with charge $q$ [27, 32]:

$$\begin{aligned}
&\mathrm{div}\ \boldsymbol{D}(\boldsymbol{r},t) = 4\pi\rho(\boldsymbol{r},t), \qquad \mathrm{div}\ \boldsymbol{B}(\boldsymbol{r},t) = 0\ ,\\
&\mathrm{rot}\ \boldsymbol{B}(\boldsymbol{r},t) = \frac{1}{c}\frac{\partial \boldsymbol{D}(\boldsymbol{r},t)}{\partial t} + \frac{4\pi}{c}\boldsymbol{j}(\boldsymbol{r},t), \qquad \mathrm{rot}\ \boldsymbol{E}(\boldsymbol{r},t) = -\frac{1}{c}\frac{\partial \boldsymbol{H}(\boldsymbol{r},t)}{\partial t}\ ,\\
&\boldsymbol{j}(\boldsymbol{r},t) = \boldsymbol{v}_0(t)\rho(\boldsymbol{r},t),\ \ \rho(\boldsymbol{r},t) = q\delta[\boldsymbol{r}-\boldsymbol{r}_0(t)]\ ;
\end{aligned} \tag{2.1}$$

$$\frac{\mathrm{d}}{\mathrm{d}t}\frac{m\boldsymbol{v}_0}{\sqrt{1-v_0^2/c^2}} = q\left\{\boldsymbol{E} + \boldsymbol{E}_0 + \frac{1}{c}\left[\boldsymbol{v}_0(\boldsymbol{H}+\boldsymbol{H}_0)\right]\right\};\ \ \boldsymbol{v}_0(t) = \frac{\mathrm{d}}{\mathrm{d}t}\boldsymbol{r}_0(t)\ . \tag{2.2}$$

Here $\boldsymbol{E}(\boldsymbol{r},t)$ and $\boldsymbol{H}(\boldsymbol{r},t)$ are the strengths of the electric and magnetic fields of the particle, respectively, and $\boldsymbol{E}_0(\boldsymbol{r},t)$ and $\boldsymbol{H}_0(\boldsymbol{r},t)$ are the strengths of the electric and magnetic external fields, respectively, which are assumed to be constant during the transmission of the charged particle through the crystal.

Equations (2.1) and (2.2) are complemented by constitutive equations, which define induction vectors of the fields in the crystal ($i,j = 1,2,3$):

$$\begin{aligned}
D_i(\boldsymbol{r},t) &= \int_{-\infty}^{t}\mathrm{d}t'\int\mathrm{d}\boldsymbol{r}'\sum_j \epsilon_{\mathrm{ij}}(t-t',\boldsymbol{r},\boldsymbol{r}')E_j(\boldsymbol{r}',t')\ ,\\
B_i(\boldsymbol{r},t) &= \int_{-\infty}^{t}\mathrm{d}t'\int\mathrm{d}\boldsymbol{r}'\sum_j \mu_{\mathrm{ij}}(t-t',\boldsymbol{r},\boldsymbol{r}')H_j(\boldsymbol{r}',t')\ ,
\end{aligned} \tag{2.3}$$

where $\epsilon_{\rm ij}$ and $\mu_{\rm ij}$ are the tensors of dielectric and magnetic permittivity for the crystal, respectively. Below only the non-magnetic crystals are considered, i.e. $\boldsymbol{B} = \boldsymbol{H},\ \boldsymbol{H}_0 = 0$. In general, some components of the tensors $\epsilon_{\rm ij}$ and $\mu_{\rm ij}$ are interdependent and, therefore, the microscopic interaction of radiation with the medium is expressed via the tensor of dielectric permittivity [32].

The Lorentz force in the motion equation (2.2) is defined by both external and radiation fields. However, restricting ourselves to spontaneous processes without collective (laser) effects, the particle energy loss for radiation yield is assumed to be negligible, and condition (1.23) allows us to neglect a recoil effect. Thus, the function $\boldsymbol{E}(\boldsymbol{r}, t)$ in (2.2) can be dropped, and then the periodic structure of the crystal takes part in the formation of the radiation field for two reasons: (i) the trajectory of the charged particle is changed because of direct interaction between the particle and electric field $\boldsymbol{E}_0(\boldsymbol{r}, t)$ of atoms, and (ii) polarization of the crystal, described by the dielectric permittivity $\epsilon_{\rm ij}(t - t', \boldsymbol{r}, \boldsymbol{r}')$. The former reason leads to well-known orientational effects, for instance, coherent bremsstrahlung, particle channelling and channelling radiation (see [2, 5, 6, 8, 15, 23, 25, 33]).

In the majority of the above mentioned works, the polarization of the medium has been partially taken into account; yet a crystal is considered to be a homogeneous medium. In microscopical theory [22, 27] of interaction between X-rays and matter, however, the periodic crystal structure makes an essential change of the spatial dispersion of dielectric permittivity, in contrast to the homogeneous medium.

In particular, the tensor of dielectric permittivity in the constitutive equation (2.3) becomes a periodic function of coordinates due to its dependence on the periodic electron density of the crystal. Thus, $\epsilon_{\rm ij}$ depends not only on the difference $\boldsymbol{r} - \boldsymbol{r}'$, as for a homogeneous medium, but on each coordinate separately.

Taking into consideration these factors, the constitutive equation (2.3) for the Fourier components of the electromagnetic field in a crystal [22] is written as

$$
\begin{aligned}
D_i(\boldsymbol{k}, \omega) &= \sum_{\boldsymbol{g}} \sum_{j} \epsilon_{\rm ij}(\boldsymbol{k}, \boldsymbol{k} + \boldsymbol{g}, \omega) E_j(\boldsymbol{k} + \boldsymbol{g}, \omega) \, , \\
\boldsymbol{E}(\boldsymbol{r}, t) &= \int {\rm d}\omega \int {\rm d}\boldsymbol{k} \boldsymbol{E}(\boldsymbol{k}, \omega) {\rm e}^{{\rm i}(\boldsymbol{kr} - \omega t)} \, , \\
\boldsymbol{D}(\boldsymbol{r}, t) &= \int {\rm d}\omega \int {\rm d}\boldsymbol{k} \boldsymbol{D}(\boldsymbol{k}, \omega) {\rm e}^{{\rm i}(\boldsymbol{kr} - \omega t)} \, , \\
\epsilon_{\rm ij}(t, \boldsymbol{r}, \boldsymbol{r}') &= \sum_{\boldsymbol{g}} \int {\rm d}\omega \int {\rm d}\boldsymbol{k} \; \epsilon_{\rm ij}(\boldsymbol{k}, \boldsymbol{k} + \boldsymbol{g}, \omega) {\rm e}^{{\rm i}\boldsymbol{k}(\boldsymbol{r} - \boldsymbol{r}') - {\rm i}\omega t} {\rm e}^{-{\rm i}\boldsymbol{g}\boldsymbol{r}'} \, .
\end{aligned}
\tag{2.4}
$$

In (2.4), the summation is performed over the whole set of reciprocal lattice vectors $\boldsymbol{g}$, and the Fourier components of dielectric permittivity are not phenomenological values but are calculated from the averaging of the induction field over the equilibrium quantum state of the crystal electron subsystem [22, 27]. Finally, the tensor $\epsilon_{\mathrm{ij}}(\boldsymbol{k}, \boldsymbol{k}+\boldsymbol{g}, \omega)$ is expressed through the amplitude of elastic coherent scattering of photons on atoms and nuclei of the crystal (see Appendix A.1). Comparing the constitutive equation (2.4) with the equation for the spatial dispersion in a homogeneous medium,

$$D_i(\boldsymbol{k}, \omega) = \sum_j \epsilon_{\mathrm{ij}}(\boldsymbol{k}, \omega) E_j(\boldsymbol{k}, \omega) \,,$$

$$\epsilon_{\mathrm{ij}}(t, \boldsymbol{r}-\boldsymbol{r}') = \int \mathrm{d}\omega \int \mathrm{d}\boldsymbol{k}\ \epsilon_{\mathrm{ij}}(\boldsymbol{k}, \omega) \mathrm{e}^{\mathrm{i}\boldsymbol{k}(\boldsymbol{r}-\boldsymbol{r}')-\mathrm{i}\omega t} \,, \tag{2.5}$$

the electromagnetic wave with the wave vector $\boldsymbol{k}$ evidently induces in the crystal a set of diffracted waves with the wave vectors $\boldsymbol{k}+\boldsymbol{g}$, contrary to a homogeneous medium. Using a language of quantum optics, the photon with the wave vector $\boldsymbol{k}$ is parametrically transformed into wave packet with the wave vectors $\boldsymbol{k}+\boldsymbol{g}$ (see Sect. 1.3). This transformation occurs in the framework of linear electrodynamics due to the momentum transfer to crystallographic lattice, in contrast to the optical wavelength range, where parametric light transformation occurs due to non-linear polarizability of the medium for interacting photons [34].

The constitutive equations (2.4) are essential for the X-ray wavelength range and make a basis for the X-ray dynamical diffraction theory, which has been intensively developed since 1930s and is widely used now in applied science [4]. In most of the studies dedicated to electromagnetic radiation from charged particles in crystals [2, 5, 6, 15, 23, 25], the spatial dispersion is taken into account in the form of (2.5) and, consequently, the results obtained are inadequate for X-ray wavelengths. The expression for the spatial dispersion (2.4) was used for the first time for calculation of electromagnetic radiation in our works [7, 10], and then comprehensively described in [8, 17].

Another technical aspect is important for calculation of PXR spectra. In real experiments, the crystal where the radiation is formed has a finite thickness. Therefore, boundary conditions at the crystal–vacuum interface have to be taken into account to solve (2.1)–(2.3). In works devoted to radiation from charged particles, dealing both with homogeneous constitutive equations [2, 5, 6, 15, 23, 25] and (2.3), which include diffraction [19], the following approach is used for boundary conditions. Maxwell's equations (2.1) for electromagnetic fields at the constant current are reduced [27] to a linear differential equation of second order for $\boldsymbol{E}(\boldsymbol{r}, t)$. The general solution for this equation is constructed as a linear combination of solutions $\boldsymbol{E}^{(1,2)}$ for homogeneous equations and a particular solution $\boldsymbol{E}^{(h)}$ for the inhomogeneous equation:

$$E_i(\boldsymbol{r}, t) = C_i^{(1)} E_i^{(1)}(\boldsymbol{r}, t) + C_i^{(2)} E_i^{(2)}(\boldsymbol{r}, t) + E_i^{(h)}(\boldsymbol{r}, t) \,. \tag{2.6}$$

The electromagnetic field $E_i(\boldsymbol{r}, t)$ is used to satisfy the regular boundary conditions of electrodynamics, which result in the system of linear inhomogeneous equations for coefficients $C_i^{(1,2)}$. The solutions for these equations, being expressed through $E_i^{(h)}(\boldsymbol{r},t)$, define the intensity of radiation at a large distance from the crystal [19]. Thus, the construction of a particular solution for inhomogeneous equations (2.1) is a principal yet difficult step for taking into account the constitutive equations (2.3) in crystals at arbitrary experimental geometry [13].

The work [11] (see also [8,17]) proposes an essentially simpler method for calculation of the radiation intensity from a charged particle moving along an arbitrary trajectory $\boldsymbol{r}_0(t)$ in a medium with any dispersion law. This method allows us to calculate radiation spectra in a crystal of finite length, using only the solution for homogeneous Maxwell's equations, which are known from the dynamical diffraction theory for any experimental geometry [4]. Excluding the magnetic field from (2.1),

$$\mathrm{rot}\ \mathrm{rot}\ \boldsymbol{E}(\boldsymbol{r},t)+\frac{1}{c^2}\frac{\partial^2\boldsymbol{D}(\boldsymbol{r},t)}{\partial t^2}=-\frac{4\pi}{c^2}\frac{\partial}{\partial t}\boldsymbol{j}(\boldsymbol{r},t)\ , \tag{2.7}$$

and changing to the Fourier space of a temporal variable, we arrive at

$$\begin{aligned}\mathrm{rot}\ \mathrm{rot}\ \boldsymbol{E}(\boldsymbol{r},\omega)-\frac{\omega^2}{c^2}\boldsymbol{D}(\boldsymbol{r},\omega)&=\mathrm{i}\omega\frac{4\pi}{c^2}\boldsymbol{j}(\boldsymbol{r},\omega)\ ,\\ \boldsymbol{j}(\boldsymbol{r},\omega)&=\frac{1}{2\pi}\int \mathrm{d}t\ q\boldsymbol{v}_0(t)\delta[\boldsymbol{r}-\boldsymbol{r}_0(t)]\mathrm{e}^{\mathrm{i}\omega t}\ .\end{aligned} \tag{2.8}$$

The Green function can be defined [28] on the left-hand side of (2.8):

$$\begin{aligned}\varepsilon_{\alpha\beta\gamma}\varepsilon_{\gamma\mu\nu}\frac{\partial^2}{\partial x_\beta\partial x_\mu}G_{\nu\lambda}(\boldsymbol{r},\boldsymbol{r}',\omega)-\frac{\omega^2}{c^2}\int \mathrm{d}\boldsymbol{r}_1\epsilon_{\alpha\beta}(\boldsymbol{r},\boldsymbol{r}_1,\omega)G_{\beta\lambda}(\boldsymbol{r}_1,\boldsymbol{r}',\omega)\\ =\delta_{\alpha\nu}\delta(\boldsymbol{r}-\boldsymbol{r}')\ ,\end{aligned} \tag{2.9}$$

where $\varepsilon_{\alpha\beta\gamma}$ is a Levi-Civita tensor [28]. The indices in (2.9) take the values $\alpha,\beta,\gamma,\mu,\nu=1,2,3$ and the summation is performed over internal indices. Considering the spontaneous radiation only, which vanishes if the current is absent, the electric field $\boldsymbol{E}(\boldsymbol{r},\omega)$ is written as

$$E_\alpha(\boldsymbol{r},\omega)=\mathrm{i}\omega\frac{4\pi}{c^2}\int \mathrm{d}\boldsymbol{r}'\ G_{\alpha\beta}(\boldsymbol{r},\boldsymbol{r}',\omega)j_\beta(\boldsymbol{r}',\omega)\ . \tag{2.10}$$

The radiation field is then delivered by the solution $\boldsymbol{E}(\boldsymbol{r},\omega)$ at a large distance from the charge, i.e. the Green function in (2.10) has to be taken in the limit $r\gg r'$. The derivation of this asymptotic for an arbitrary tensor of dielectric permittivity is given in [8,17] and described in detail in Appendix A.2. The resulting approximation is

$$G_{\alpha\beta}(\boldsymbol{r},\boldsymbol{r}',\omega)\approx\frac{\mathrm{e}^{\mathrm{i}kr}}{4\pi r}\sum_{s=1,2}e_\alpha^{(s)}E_{\boldsymbol{k}\beta s}^{(-)*}(\boldsymbol{r}',\omega),\quad \boldsymbol{k}=\frac{\omega}{c}\frac{\boldsymbol{r}}{r}\ , \tag{2.11}$$

where $\boldsymbol{e}^{(s)}$ are the polarization vectors, and $E^{(-)*}_{\boldsymbol{k}\beta s}(\boldsymbol{r}',\omega)$ are solutions of the homogeneous Maxwell's equation

$$(\text{rot rot } \boldsymbol{E}^{(-)}_{\boldsymbol{k}s}(\boldsymbol{r},\omega))_\alpha - \frac{\omega^2}{c^2}\int \mathrm{d}\boldsymbol{r}_1\, \epsilon^*_{\alpha\beta}(\boldsymbol{r},\boldsymbol{r}_1,\omega)E^{(-)}_{\boldsymbol{k}\beta s}(\boldsymbol{r}_1,\omega) = 0\,, \quad (2.12)$$

which satisfies the following asymptotic boundary condition:

$$\boldsymbol{E}^{(-)}_{\boldsymbol{k}s}(\boldsymbol{r},\omega) \approx \boldsymbol{e}^{(s)}\mathrm{e}^{\mathrm{i}\boldsymbol{k}\boldsymbol{r}} + \boldsymbol{f}_{\mathrm{s}}\frac{\mathrm{e}^{-\mathrm{i}kr}}{r}, \quad r\to\infty\,, \quad (2.13)$$

with amplitudes $\boldsymbol{f}_{\mathrm{s}}$ independent of $r$.

Thus, the solution containing the asymptotically convergent spherical wave is used for describing the radiation produced inside the crystal. This is contrary to the scattering (diffraction) of the external electromagnetic wave from the same medium, when the electromagnetic field $\boldsymbol{E}^{(+)}_{\boldsymbol{k}s}(\boldsymbol{r},\omega)$ includes a divergent spherical wave at an infinite distance. Both these solutions are simply connected with each other by the equation (Appendix A.2)

$$(\boldsymbol{E}^{(-)}_{\boldsymbol{k}s}(\boldsymbol{r},\omega))^* = \boldsymbol{E}^{(+)}_{-\boldsymbol{k}s}\,, \quad (2.14)$$

which corresponds to the well-known optics reciprocity theorem [16].

To calculate the spectral density of the radiation energy $W_{\boldsymbol{n}\omega}$ normalized to the spatial angle around the observation vector $\boldsymbol{n} = \boldsymbol{k}/k$, asymptotic (2.11) for the Green function is substituted into (2.10). Using the Umov–Poynting vector for derivation of the energy density [20],

$$W_{\boldsymbol{n}\omega} = \frac{cr^2}{4\pi^2}|\boldsymbol{E}(\boldsymbol{r},\omega)|^2, \quad (2.15)$$

and inserting (2.10) with asymptotic (2.11) and (2.8) into (2.15), two equivalent expressions for the energy density [8,17] are found:

$$\begin{aligned} W^{(s)}_{\boldsymbol{n}\omega} &= \frac{q^2\omega^2}{4\pi^2c^3}\left|\int_{-\infty}^{\infty}\mathrm{d}t\ \mathrm{e}^{\mathrm{i}\omega t}\boldsymbol{v}_0(t)\boldsymbol{E}^{(-)*}_{\boldsymbol{k}s}(\boldsymbol{r}_0(t),\omega)\right|^2 \\ &= \frac{q^2\omega^2}{4\pi^2c^3}\left|\int_{-\infty}^{\infty}\mathrm{d}t\ \mathrm{e}^{\mathrm{i}\omega t}\boldsymbol{v}_0(t)\boldsymbol{E}^{(+)}_{-\boldsymbol{k}s}(\boldsymbol{r}_0(t),\omega)\right|^2 . \end{aligned} \quad (2.16)$$

This formula can be used for calculation of the intensity of spontaneous radiation from a charged particle moving in the media with an arbitrary dispersion. Equation (2.16) is the general form of the classic electrodynamics formula, derived from the spectral expansion of a retarded potential [26]:

$$W^{(s)}_{\boldsymbol{n}\omega} = \frac{q^2\omega^2}{4\pi^2c^3}\left|\int_{-\infty}^{\infty}\mathrm{d}t\ (\boldsymbol{v}_0(t)\boldsymbol{e}_s)\mathrm{e}^{\mathrm{i}\omega t - \mathrm{i}\boldsymbol{k}\boldsymbol{r}_0(t)}\right|^2 . \quad (2.17)$$

Instead of using the intensity distribution (2.16) normalized to one particle, the spectral–angular density of the photons emitted by the beam of charged particles of current $J$ in unit time period is often used:

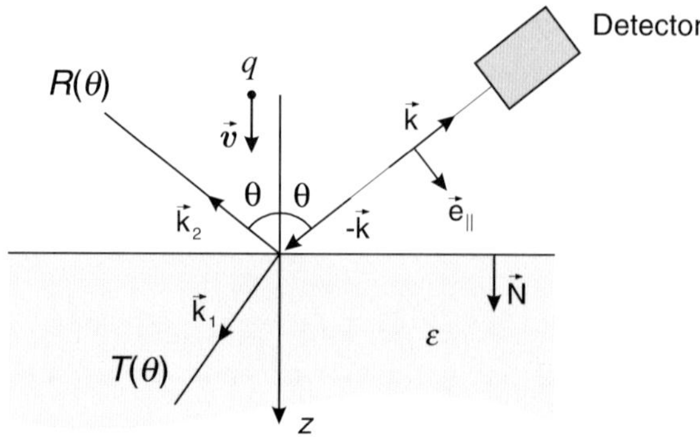


**Fig. 2.1.** Wave fields for the calculation of the transition radiation intensity

$$\frac{\partial^2 N}{\partial \boldsymbol{n} \partial \omega} = \frac{W^{(s)}_{\boldsymbol{n}\omega}}{\hbar\omega} \frac{J}{q} \,. \tag{2.18}$$

To illustrate the above-described method, the intensity of transition radiation [21] from a constantly moving charged particle crossing the boundary between two different media is calculated below.

Let us consider a semi-infinite homogeneous medium with a surface at $z = 0$, where the $z$-axis is parallel to the inward normal $\boldsymbol{N}$ to the surface. The medium possesses a dielectric permittivity $\epsilon$ and the velocity of a charged particle is parallel to the surface normal $\boldsymbol{v} \| \boldsymbol{N}$ (Fig. 2.1). If a detector is placed in vacuum at an angle $\theta$ to the velocity of the particle, the wave vector $\boldsymbol{k}$ has components $k_x = -k \sin\theta,\ k_y = 0, k_z = -k\cos\theta,\ k = \omega/c$, in the chosen coordinate system in Fig. 2.1. According to (2.16), the radiation intensity can be calculated using the solution of homogeneous Maxwell's equations, which describe the reflection and refraction of a plane wave with unit amplitude, wave vector $\boldsymbol{k}_0 = -\boldsymbol{k}$ and polarization $\boldsymbol{e}_{\mathrm{s}}$ at the boundary. This solution follows from Fresnel formulas [27]. Because integral (2.16) contains the term $(\boldsymbol{v}\boldsymbol{e}_{\mathrm{s}})$, the radiation is polarized within the plane $\boldsymbol{N}, \boldsymbol{k}_0$ and the intensity is proportional to $E_z$, which is [27]

$$\begin{aligned} E_z(x,z) &= -\sin\theta \mathrm{e}^{\mathrm{i}k\sin\theta x}\Psi(z) \,, \\ \Psi(z) = \Psi_1(z) &= \mathrm{e}^{\mathrm{i}k\cos\theta z} + R(\theta)\mathrm{e}^{-\mathrm{i}k\cos\theta z} , \qquad z < 0 \,, \\ \Psi(z) = \Psi_2(z) &= T(\theta)\mathrm{e}^{\mathrm{i}k_{1z}z} , \qquad k_{1z} = k\sqrt{\epsilon - \sin^2\theta}, \quad z > 0 \,. \end{aligned} \tag{2.19}$$

The coefficients of reflection $R$ and transmission $T$ are

$$\begin{aligned} R(\theta) &= \frac{\epsilon\cos\theta - \sqrt{\epsilon - \sin^2\theta}}{\epsilon\cos\theta + \sqrt{\epsilon - \sin^2\theta}} \,, \\ T(\theta) &= \frac{2\cos\theta}{\epsilon\cos\theta + \sqrt{\epsilon - \sin^2\theta}} \,. \end{aligned} \tag{2.20}$$

Substituting (2.19) into (2.16) results in the integral ($\beta = v/c$)

$$I = \int_{-\infty}^{0} \mathrm{d}t\ v\Psi_1(vt)\mathrm{e}^{\mathrm{i}\omega t} + \int_{0}^{\infty} \mathrm{d}t\ v\Psi_2(vt)\mathrm{e}^{\mathrm{i}\omega t}$$
$$= \frac{\mathrm{i}v\sin\theta}{\omega}\left[\frac{1}{1+\beta\cos\theta} + \frac{R}{1-\beta\cos\theta} - \frac{T}{1+\beta\sqrt{\epsilon-\sin^2\theta}}\right] . \quad (2.21)$$

After some algebraical transformations of (2.20) and (2.21), (2.16) for the intensity of transition radiation can be rewritten in the well-known form [20]:

$$W_{\boldsymbol{n}\omega} = \frac{q^2v^2\sin^2\theta\cos^2\theta}{\pi^2c^3(1-\beta^2\cos^2\theta)^2}|F(\omega,\theta)|^2 ,$$
$$F(\omega,\theta) = \frac{(\epsilon-1)(1-\beta^2+\beta\sqrt{\epsilon-\sin^2\theta})}{(\epsilon\cos\theta+\sqrt{\epsilon-\sin^2\theta})(1+\beta\sqrt{\epsilon-\sin^2\theta})} . \quad (2.22)$$

## 2.2 PXR from Relativistic Electrons in Thin Crystals

The specific features of PXR become apparent in the case of transmission of a charged particle through a thin crystal (monocrystalline film) of thickness $L$, which is less than the extinction length $L_{\mathrm{ext}}$ in (1.24). In this case [27], the solution $\boldsymbol{E}^{(+)}_{\boldsymbol{k}s}(\boldsymbol{r},\omega)$ for homogeneous Maxwell's equations, which describes the diffraction of X-rays, can be found using a perturbation theory developed on the crystal polarizability $\chi_{\mathrm{ij}} = \epsilon_{\mathrm{ij}} - 1$. According to (2.16), the PXR spectrum can be calculated with equivalent accuracy by simple analytical formulas.

The coherent interaction of X-rays with the crystal, which defines the X-ray polarizability $\chi_{\mathrm{ij}}$, includes Compton scattering of photons on electrons and resonant scattering on atomic and nuclear transitions. Both these modes are utilized in different applications and discussed later.

The X-ray polarizibility of the crystal due to scattering of radiation on electrons of atoms is calculated, for example, in [22, 27]:

$$\chi_{\mathrm{ij}} = -\frac{4\pi e^2}{m\omega^2}\delta_{\mathrm{ij}}n(\boldsymbol{r}) , \quad (2.23)$$

where $e$ and $m$ are the charge and mass of the electron, respectively, and $n(\boldsymbol{r})$ is the electron density, which is calculated from the averaging of the density operator over the quantum state of the crystal. This electron density does not follow from averaging over physically small volume, as in macroscopical theory. It is a periodic function of coordinates and can be expanded in Fourier series on reciprocal lattice vectors of a crystal:

$$n(\boldsymbol{r}) = \sum_{\boldsymbol{g}} n_{\boldsymbol{g}}\mathrm{e}^{\mathrm{i}\boldsymbol{g}\boldsymbol{r}} , \quad (2.24)$$

where the expansion coefficients of the density $n_{\boldsymbol{g}}$ depend on the volume and scattering properties of the crystallographic unit cell. Neglecting the absorption in a thin crystal, these coefficients are related to the Fourier components of X-ray polarizabilities $\chi_{\boldsymbol{g}}$ as (Appendix A.1)

$$\chi_{\boldsymbol{g}} = -\frac{4\pi e^2}{m\omega^2} n_{\boldsymbol{g}} \,.$$

Within the limits of the Born approximation for $\chi_{\boldsymbol{g}}$, the solution for (2.12) can be found as [27]

$$\begin{aligned} \boldsymbol{E}^{(+)}_{-\boldsymbol{k}s}(\boldsymbol{r},\omega) &= \boldsymbol{e}_{\mathrm{s}} \mathrm{e}^{-\mathrm{i}\boldsymbol{k}\boldsymbol{r}} + \boldsymbol{E}^{(sc)}_{-\boldsymbol{k}s}(\boldsymbol{r},\omega) \,, \\ \boldsymbol{E}^{(sc)}_{-\boldsymbol{k}s} &= \frac{e^2}{m\omega^2} \operatorname{rot}\operatorname{rot} \boldsymbol{e}_{\mathrm{s}} \int_V \mathrm{d}\boldsymbol{r}' \frac{\mathrm{e}^{\mathrm{i}k|\boldsymbol{r}-\boldsymbol{r}'|}}{|\boldsymbol{r}-\boldsymbol{r}'|} n(\boldsymbol{r}') \mathrm{e}^{-\mathrm{i}\boldsymbol{k}\boldsymbol{r}'} \,. \end{aligned} \tag{2.25}$$

Here $\boldsymbol{e}_{\mathrm{s}}$ is the polarization and the vector $\boldsymbol{k} = k\boldsymbol{n} \equiv \frac{\omega}{c}\boldsymbol{n}$, $\boldsymbol{n} = \frac{\boldsymbol{r}}{r}$, points to the detector position. The integration is performed over the crystal volume $V$.

The motion law $\boldsymbol{r}(t) = \boldsymbol{v}t$, corresponding to the constantly moving charged particle, has to be used for calculation of the PXR intensity from (2.17). The contribution of unperturbative part of the wave field (2.25), which is proportional to

$$\int_{-\infty}^{-\infty} \mathrm{d}t\; \mathrm{e}^{\mathrm{i}(\omega - \boldsymbol{k}\boldsymbol{v})t} \,,$$

is equal to zero because a free charge does not radiate. The scattered field $\boldsymbol{E}^{(sc)}_{-\boldsymbol{k}s}$ can be written using the integral representation of the Green function and expression (2.24) for the electron density:

$$\boldsymbol{E}^{(sc)}_{-\boldsymbol{k}s} = -\sum_{\boldsymbol{g}} \chi_{\boldsymbol{g}} \int_V \mathrm{d}\boldsymbol{r}' \int \frac{\mathrm{d}\boldsymbol{p}}{(2\pi)^3} \frac{[\boldsymbol{p}[\boldsymbol{p}\boldsymbol{e}_{\mathrm{s}}]]}{p^2 - \omega^2} \mathrm{e}^{\mathrm{i}\boldsymbol{p}(\boldsymbol{r}-\boldsymbol{r}')} \mathrm{e}^{\mathrm{i}(\boldsymbol{g}-\boldsymbol{k})\boldsymbol{r}'} \,. \tag{2.26}$$

Substituting (2.26) into (2.17), the PXR spectral–angular distribution is

$$\begin{aligned} W^{(s)}_{\boldsymbol{n}\omega} = \frac{q^2\omega^2}{4\pi^2 c^3} \Bigg| \sum_{\boldsymbol{g}} \chi_{\boldsymbol{g}} \int_V \mathrm{d}\boldsymbol{r}' \int \frac{\mathrm{d}\boldsymbol{p}}{(2\pi)^3} \frac{(\boldsymbol{v}[\boldsymbol{p}[\boldsymbol{p}\boldsymbol{e}_{\mathrm{s}}]])}{p^2 - \omega^2} \\ \int_V \mathrm{d}\boldsymbol{r}' \mathrm{e}^{\mathrm{i}(\boldsymbol{p}+\boldsymbol{g}-\boldsymbol{k})\boldsymbol{r}'} \int_{-\infty}^{-\infty} \mathrm{d}t\; \mathrm{e}^{\mathrm{i}(\omega - \boldsymbol{p}\boldsymbol{v})t} \Bigg|^2 \,. \end{aligned} \tag{2.27}$$

If the $z$-axis is chosen along the velocity $\boldsymbol{v}$ and the cross-section of the crystal $S$ is large enough, the integration over $t$ and $\boldsymbol{r}'_\perp$ is reduced to $\delta$-functions, which cancel the integration over $\boldsymbol{p}$ at the point

$$\boldsymbol{p} = \boldsymbol{p}_g; \qquad p_{gz} = \frac{\omega}{v}, \qquad \boldsymbol{p}_{g\perp} = (\boldsymbol{k} - \boldsymbol{g})_\perp \,. \tag{2.28}$$

The PXR intensity then is

$$W^{(s)}_{\boldsymbol{n}\omega} = \frac{q^2\omega^2}{4\pi^2c^3}\left|\sum_{\boldsymbol{g}}\chi_{\boldsymbol{g}}\frac{(\boldsymbol{v}[\boldsymbol{p}_g[\boldsymbol{p}_g\boldsymbol{e}_{\mathrm{s}}]])}{(\boldsymbol{k}-\boldsymbol{g})^2_\perp+\frac{\omega^2}{v^2}(1-\frac{v^2}{c^2})}\int_0^L\frac{\mathrm{d}z}{v}\mathrm{e}^{\mathrm{i}(\frac{\omega}{v}+g_z-k_z)z}\right|^2 . \quad (2.29)$$

This expression demonstrates that in contrast to the Vavilov–Cherenkov radiation, the PXR by its nature is not a threshold phenomenon, because $W^{(s)}_{\boldsymbol{n}\omega}$ is non-zero at any energy $E$ of the charged particle. The PXR intensity increases with the particle energy in a relativistic case when the following condition is fulfilled:

$$\left(1-\frac{v^2}{c^2}\right)=\left(\frac{mc^2}{E}\right)^2=\gamma^{-2}\ll 1 ,$$

and the radiation is concentrated in the vicinity of directions:

$$|(\boldsymbol{k}-\boldsymbol{g})_\perp| \le \frac{\omega}{v\gamma} . \quad (2.30)$$

The integral in (2.29) has resonant behaviour in this case: For arbitrary wave vectors of an emitted photon, it is of order $k^{-1}\sim\lambda$ and reaches its maximum close to $L\gg\lambda$ under the condition

$$\nu=\frac{1}{2}\left|\frac{\omega}{v}+g_z-k_z\right|\le\frac{1}{L}, \qquad L<L_{\mathrm{ext}} . \quad (2.31)$$

For relativistic particles, the resonances for different vectors $\boldsymbol{g}$ in (2.29) do not interfere; therefore the PXR intensity equals

$$W^{(s)}_{\boldsymbol{n}\omega}=\sum_{\boldsymbol{g}}W^{(s)}_{\boldsymbol{g}} ;$$

$$W^{(s)}_{\boldsymbol{g}} = \frac{q^2\omega^2}{4\pi^2c^3}|\chi_{\boldsymbol{g}}|^2\frac{(\boldsymbol{v}[\boldsymbol{p}_g[\boldsymbol{p}_g\boldsymbol{e}_{\mathrm{s}}]])^2}{\left[(\boldsymbol{k}-\boldsymbol{g})^2_\perp+\frac{\omega^2}{v^2\gamma^2}\right]^2}\frac{\sin^2\nu L}{v^2\nu^2} . \quad (2.32)$$

Thus, the PXR spectral–angular distribution from a relativistic particle is represented by a diffractional set of peaks (reflections), corresponding to different reciprocal lattice vectors of the crystal. The radiation emitted in each energy peak is highly monochromatic. The wave vectors of the photons within the reflection are allocated near the vector defined by conditions (2.30), (2.31) and $k=\omega/c$:

$$\boldsymbol{k}_{\mathrm{B}}=\frac{\omega_{\mathrm{B}}\boldsymbol{v}}{v^2}+\boldsymbol{g}, \qquad \omega_{\mathrm{B}}=-\frac{g^2}{2g_z}=\frac{g}{2\sin\theta_{\mathrm{B}}} , \quad (2.33)$$

where the parameter $\theta_{\mathrm{B}}$ is the angle between $\boldsymbol{v}$ and the crystallographic planes, corresponding to the reciprocal lattice vector $\boldsymbol{g}$ (Fig. 2.3). PXR photons in the reflections from the plane sets with Miller indices multiple to the minimal set $\{hkl\}$ have multiple frequencies and are concentrated near the same spatial direction. For the plane sets with reciprocal vectors

$$(\boldsymbol{k}_\mathrm{B} + \boldsymbol{u})^2 = k^2 = \frac{\omega^2}{c^2}, \tag{2.40}$$

$$2(\boldsymbol{k}_\mathrm{B}\boldsymbol{u}) = 2\frac{\omega_\mathrm{B}}{v}(u_x \sin 2\theta_\mathrm{B} + u_z \cos 2\theta_\mathrm{B}) \approx \frac{\omega^2}{c^2} - \frac{\omega_\mathrm{B}^2}{v^2},$$

$$u_z \approx \frac{\omega - \omega_\mathrm{B}}{c \cos 2\theta_\mathrm{B}} - \theta_x \frac{\omega_\mathrm{B}}{c} \tan 2\theta_\mathrm{B}. \tag{2.41}$$

Let us assume that the polarization vector $\boldsymbol{e}_\mathrm{s}$, accepted by a detector, makes an angle $\varphi$ with the diffraction planes. We consider below the radiation from electrons, i.e. $q = e$ in (2.32), and use (2.18) to analyse the spectral–angular distribution of photons, emitted into the set of reflections $\{hkl\}$. The cross-section of PXR then is (with the accuracy $o(\gamma^{-2})$)

$$\frac{\partial^3 N^{(s)}_{\{hkl\}}}{\partial\theta_x \partial\theta_y \partial\omega} = \frac{\alpha}{2\pi}\frac{J}{e}\sum_{m=1}^{\infty} \frac{(\theta_x \cos 2\theta_\mathrm{B} \cos\varphi + \theta_y \sin\varphi)^2}{(\theta_x^2 + \theta_y^2 + \gamma^{-2})^2} \times |\chi_{g_m}(\omega_\mathrm{B}^{(m)})|^2 \frac{\omega_\mathrm{B}^{(m)} L}{c}$$
$$\delta\left[\frac{2\sin^2\theta_\mathrm{B}}{\cos 2\theta_\mathrm{B}}(\omega - \omega_\mathrm{B}^{(m)}) - \theta_x \omega_\mathrm{B}^{(m)} \tan 2\theta_\mathrm{B}\right], \tag{2.42}$$

where $\alpha = e^2/\hbar c \approx 1/137$ is a fine structure constant. If the photons of both polarizations are detected, (2.42) contains the sum of orthogonal polarizations:

$$\frac{\partial^3 N_{\{hkl\}}}{\partial\theta_x \partial\theta_y \partial\omega} = \frac{\alpha}{2\pi}\frac{J}{e}\sum_{m=1}^{\infty} \frac{\theta_x^2 \cos^2 2\theta_\mathrm{B} + \theta_y^2}{(\theta_x^2 + \theta_y^2 + \gamma^{-2})^2} \times |\chi_{g_m}(\omega_\mathrm{B}^{(m)})|^2 \frac{\omega_\mathrm{B}^{(m)} L}{c}$$
$$\delta\left[\frac{2\sin^2\theta_\mathrm{B}}{\cos 2\theta_\mathrm{B}}(\omega - \omega_\mathrm{B}^{(m)}) - \theta_x \omega_\mathrm{B}^{(m)} \tan 2\theta_\mathrm{B}\right]. \tag{2.43}$$

Similar to the X-ray Bragg diffraction, where the width of the diffraction peak is defined by the angular and frequency dispersion of the incident beam [27], the PXR fine structure is also determined by the dispersion of the emitted photons, both on angles and frequencies. Therefore, in the experiments on PXR two types of detectors are used. The first type is, for example, an X-ray film, where the photons of any frequency are registered and the density of angular distribution is investigated. According to (2.43), this distribution is localized on the scale $\delta\theta \sim \gamma^{-1}$; hence the following re-scaling of an angular variable is reasonable:

$$x = \gamma\theta_x, \qquad y = \gamma\theta_y,$$

which delivers the results in a form which is independent of the particle energy [18] and the reflection indices:

$$\frac{\partial^2 N_{\{hkl\}}}{\partial x \partial y} = N^{(\mathrm{tot})}_{\{hkl\}} \frac{x^2 \cos^2 2\theta_\mathrm{B} + y^2}{(x^2 + y^2 + 1)^2},$$

$$N^{(\mathrm{tot})}_{\{hkl\}} = \frac{\alpha \cos 2\theta_\mathrm{B}}{4\pi \sin^2\theta_\mathrm{B}} \frac{J}{e} \sum_{m=1}^{\infty} |\chi_{g_m}(\omega_\mathrm{B}^{(m)})|^2 \frac{\omega_\mathrm{B}^{(m)} L}{c}. \tag{2.44}$$

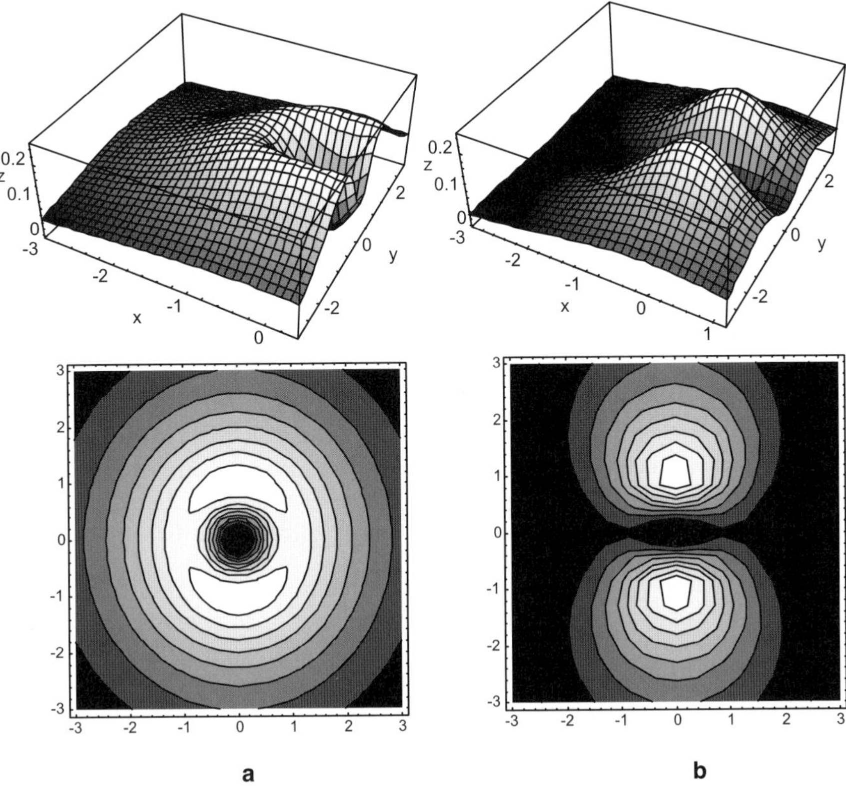


**Fig. 2.4.** The universal fine angular structure of PXR reflections, corresponding to $\theta_B = 9°$ (**a**) and $\theta_B = 30°$ (**b**)

Figure 2.4a and b show two dimensional angular distributions of the PXR intensity for $\theta_B = 9°$ and $\theta_B = 30°$, respectively. The specific features of these distributions are (i) vanishing of intensity in the direction of the vector $\boldsymbol{k}_{\{hkl\}}$ and (ii) twofold shape of pattern. Both features have been confirmed in the first PXR experiments [9].

The frequency spectrum of PXR can be studied by the second type of detectors, energy dispersive ones, which register the photons of the PXR reflection in the frequency scale. As follows from (2.43), this spectrum is a series of peaks near the frequencies $\omega_B^{(m)} = m\omega_{\{hkl\}}$, which are associated with permitted reflections $\chi_{g_m}(\omega_B^{(m)}) \neq 0$. The intensity of peaks inside the series decreases accordingly:

$$I_m \sim \omega|\chi(\omega)|^2 \sim m^{-3} \quad .$$

The spectral distribution near each peak can be derived from (2.43) by integration over the angles, and written in the universal form using a variable:

$$u = \gamma \tan\theta_{\mathrm{B}} \frac{(\omega - \omega_{\mathrm{B}}^{(m)})}{\omega_{\mathrm{B}}^{(m)}} ,$$

$$\Phi(u) = \frac{u^2(1+\cos^2 2\theta_{\mathrm{B}}) + 1}{(u^2+1)^{3/2}} ,$$

$$\frac{\partial N_{\{hkl\}}^{(m)}}{\partial u} = \frac{\alpha \cos 2\theta_{\mathrm{B}}}{8\sin^2\theta_{\mathrm{B}}} \frac{J}{e} |\chi_{g_m}(\omega_{\mathrm{B}}^{(m)})|^2 \frac{\omega_{\mathrm{B}}^{(m)} L}{c} \Phi(u) , \tag{2.45}$$

The simulated PXR spectral distributions are shown in Fig. 2.5, and experimental measurements can be performed using a detector with the high energy resolution $\Delta\omega/\omega \sim \gamma^{-1}$. The number of photons emitted in unit time and within the angular and spectral intervals of the PXR reflection is an important characteristics of the parametric X-ray radiation. This value is evaluated by integration either over $x, y$ in (2.44), or over $u$ in (2.45). Both integrals diverge logarithmically on the upper limit because the validity of (2.42) is limited in the region $\theta_x \sim \theta_y \sim \gamma^{-1} \ll 1$. To overcome this obstacle, we limit the integration area by $|u| < \xi_{\mathrm{D}} \gg 1$, which corresponds to a real angular resolution of the detector, $\theta_{\mathrm{D}}$ [18]:

$$\xi_{\mathrm{D}} \sim \gamma\theta_{\mathrm{D}} ,$$

and the full intensity of PXR photons in each spectral peak of reflection weakly depends on this instrumental parameter:

$$N_{\{hkl\}}^{(m)} \approx \frac{\alpha \cos 2\theta_{\mathrm{B}}}{4\sin^2\theta_{\mathrm{B}}} \frac{J}{e} |\chi_{g_m}(\omega_{\mathrm{B}}^{(m)})|^2 \frac{\omega_{\mathrm{B}}^{(m)} L}{c} \times\{(1+\cos^2 2\theta_{\mathrm{B}}) \ln(\xi_{\mathrm{D}} + \sqrt{\xi_{\mathrm{D}}^2 + 1}) - \cos^2 2\theta_{\mathrm{B}}\} . \tag{2.46}$$

The quasi-monochromatic beam contributing to this peak has the shape

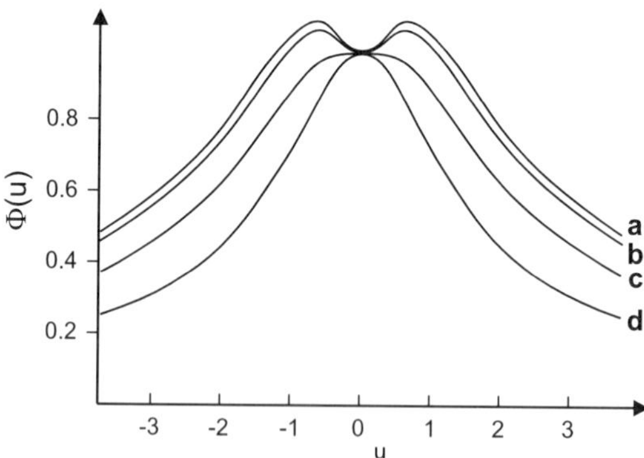


**Fig. 2.5.** The universal spectral fine structure of the PXR reflection: (**a**) $\theta_{\mathrm{B}} = 5°$, (**b**) $\theta_{\mathrm{B}} = 15°$, (**c**) $\theta_{\mathrm{B}} = 30°$, (**d**) $\theta_{\mathrm{B}} = 45°$

$$\omega^{(m)}_{\{hkl\}} = \frac{\pi mc}{d_{\{hkl\}}\sin\theta_{\mathrm{B}}}, \qquad \Delta\theta_x \approx \Delta\theta_y \approx \frac{\Delta\omega}{\omega} \approx \gamma^{-1}\,,$$
$$\frac{\omega^{(m)}_{\mathrm{B}} L}{c} \le |\chi_0|^{-1}\,. \tag{2.47}$$

The quantitative estimate of the PXR intensity in a thin crystal is possible using a parameter of brightness $I$, which is now widely accepted for characterization of synchrotron radiation (SR) [24]:

$$I = \frac{N}{\Delta\theta_x \Delta\theta_y \Delta\omega/\omega} = I_0 \frac{\text{photons}}{\text{s mrad}^2(0.1\%\,\text{bandwidth})}\,.$$

For the crystal of maximal thickness (2.47), the electron beam of energy $E \geq 0.529$ GeV, and angular and spectral divergence $\Delta\theta \approx 1$ mrad, $\Delta\omega/\omega \approx |\chi_0|$, respectively, the brightness in PXR reflection is

$$I^{(m)}_{\{hkl\}} \approx \alpha \frac{J}{e} \frac{|\chi_{g_m}(\omega^{(m)}_{\mathrm{B}})|^2}{|\chi_0|}\,. \tag{2.48}$$

To produce a parametric X-ray radiation with the energy $\hbar\omega_0 = 3$ keV from the $\{111\}$ Ge reflection (the lattice constant $a = 5.32$ Å), the angle between the photon beam and the crystallographic plane must be

$$\sin\theta_{\mathrm{B}} = \frac{\pi c\sqrt{3}}{\omega_0 a} \approx 0.68\,.$$

The crystal polarizabilities follow from known formulas (see Appendix A.1)

$$|\chi_0| \approx 2.1 \times 10^{-4}, \qquad |\chi_{\{111\}}| \approx 1.1 \times 10^{-4}\,,$$

which when substituted into (2.48) results in

$$I_{\mathrm{PXR}} \approx 1.3 \times 10^{13} J[\mathrm{A}] \frac{\text{photons}}{\text{s mrad}^2(0.1\%\,\text{bandwidth})}\,.$$

Comparing this estimate with synchrotron radiation, the same energy of photons can be reached in synchrotron by electrons of energy $E \geq 2.5$ GeV [24]:

$$I_{\mathrm{SR}} \approx 6.1 \times 10^{13} J[\mathrm{A}] \frac{\text{photons}}{\text{s mrad}^2(0.1\%\,\text{bandwidth})}\,.$$

## 2.3 Mosaicity of Crystals and Multiple Scattering of an Electron Beam

The kinematic theory of X-ray diffraction is valid not only for thin single crystals with the thickness (1.24), but also for thick mosaic polycrystalline samples [31]. These kinds of samples consist of a large number of thin monocrystalline blocks with the thickness $L_i < L_{\mathrm{ext}}$, which are twisted at a random angle relative to each other. The diffracted wave fields from different blocks are incoherent; thus the full diffracted intensity is a sum of diffraction intensities from the blocks. The X-ray intensity scattered from a single block can be calculated on the basis of the kinematic diffraction theory. The summation of intensities from different blocks is replaced by averaging over random distribution of the orientation of the normal vector $\boldsymbol{Z}$ to the diffracting plane around the vector $\boldsymbol{Z}_{\{hkl\}}$ corresponding to a perfect crystal:

$$I_{\text{tot}} = \sum_i L_i W(\boldsymbol{Z}_i) = L \int \mathrm{d}\boldsymbol{Z}\ \varphi\left(\frac{\boldsymbol{Z} - \boldsymbol{Z}_{\{hkl\}}}{\nu}\right) W(\boldsymbol{Z})\,. \tag{2.49}$$

Here $W(\boldsymbol{Z})$ is an intensity of a reflected wave, normalized to the length $L = 1$, $\varphi$ is a normalized distribution function depending on the dimensionless mosaicity parameter $\nu$. The mosaicity causes the broadening of the diffraction peak, and the full intensity is determined by the sample thickness $L = \sum L_i \gg L_{\text{ext}}$ [31].

The PXR reflections in the mosaic crystals are formed in the same way as in a single crystal. According to (2.16), they are derived from similar Maxwell's equations as for X-ray diffraction. For PXR the divergence of the velocity vector of the electron beam has to be taken into account, additionally. This divergence increases along the trajectory of the beam within the crystal due to multiple scattering on atoms. In the kinematic theory, the PXR output depends on the angle between the crystallographic planes and electron velocity. Therefore, the divergence of the electron beam and crystal mosaicity can be treated in a similar way.

The general approach for calculation of the radiation intensity in the crystal, taking into account the effects of multiple scattering and energy loss, has been considered in detail in [8]. This approach is based on the averaging of (2.16) over the trajectory and utilizes the probability density $w(\boldsymbol{r}, \boldsymbol{v}, t; \boldsymbol{r}', \boldsymbol{v}', t')$ for a particle to have the coordinate $\boldsymbol{r}$ and velocity $\boldsymbol{v}$ in time $t$, and $\boldsymbol{r}'$, $\boldsymbol{v}'$ in time $t'$ (see Appendix A.3). In the case of thin samples, the correlations between the particle states in different time moments can be neglected. If the sample is perfectly mosaic, the interference of radiation from different blocks can be ignored, too. Then the calculation of radiation from single crystalline blocks is reduced to the averaging of (2.43) over the electron velocity vector $\boldsymbol{n} = \boldsymbol{v}/v$ and normal vector $\boldsymbol{Z}$ to the scattering planes and particle energy $E$:

$$\begin{aligned}\frac{\partial^3 N_{\{hkl\}}}{\partial\theta_{x0}\partial\theta_{y0}\partial\omega} &\equiv \left\langle \frac{\partial^3 N_{\{hkl\}}}{\partial\theta_x\partial\theta_y\partial\omega} \right\rangle = \int \mathrm{d}\boldsymbol{n}\ \mathrm{d}\boldsymbol{Z}\ \mathrm{d}E\ \varphi_1\left(\frac{\boldsymbol{n} - \boldsymbol{n}_0}{\Delta\theta}\right) \\ &\times \varphi_2\left(\frac{\boldsymbol{Z} - \boldsymbol{Z}_{\{hkl\}}}{\nu}\right)\ \varphi_3\left(\frac{E - E_0}{\Delta E}\right)\ \frac{\partial^3 N_{\{hkl\}}}{\partial\theta_x\partial\theta_y\partial\omega}(\boldsymbol{n}, \boldsymbol{Z}, E)\,. \end{aligned} \tag{2.50}$$

Here $\varphi_i$ are the normalized functions of the probability distribution of the above-mentioned variables, and the parameters $\Delta\theta, \Delta E$ are root-mean-square dispersions of the velocity vector and energy, respectively. The value of $\Delta\theta$ depends both on the primary beam divergence $\Delta\theta_0$ and on the angle of multiple electron scattering in the crystal $\theta_{\rm s}(L)$, which follows from (1.26). In real experiments, this angle is essentially larger than the primary beam divergence; thus

$$\Delta\theta_L = \sqrt{(\Delta\theta_0)^2 + \theta_{\rm s}^2(L)} \approx \theta_{\rm s}(L) \, .$$

The crystallographic structure of a sample is not taken into consideration in multiple scattering calculations; therefore $\theta_{\rm s}(L)$ does not depend on mosaicity and is determined by the total sample thickness.

To analyse (2.50), new integration variables are introduced as follows:

$$\epsilon = E - E_0, \qquad \boldsymbol{s} = (\boldsymbol{n} - \boldsymbol{n}_0), \qquad \boldsymbol{z} = (\boldsymbol{Z} - \boldsymbol{Z}_{\{hkl\}});$$

$$\boldsymbol{s} \perp \boldsymbol{n}_0, \qquad \boldsymbol{z} \perp \boldsymbol{Z}_{\{hkl\}}, \qquad \boldsymbol{n}_0 = \frac{\boldsymbol{v}_0}{v_0}, \qquad |\epsilon| \ll E_0, \qquad |\boldsymbol{s}| \ll 1 \, ,$$

and the angular variables $\theta_{x,y}$ in (2.40) are represented in a covariant form:

$$\theta_x = \frac{(\boldsymbol{ug}) - (\boldsymbol{un})(\boldsymbol{gn})}{\omega_{\rm B} g}; \qquad \theta_y = \frac{(\boldsymbol{u}[\boldsymbol{gn}])}{\omega_{\rm B} g} \, .$$

Then with the accuracy $o(\theta_{\rm s}^2, \, \gamma^{-2})$ ,

$$\theta_x \approx \theta_{x0} + \cos\theta_{\rm B} q_x, \quad \theta_y \approx \theta_{y0} + \cos\theta_{\rm B} q_y, \quad v \approx v_0 + \frac{\epsilon}{E_0}\gamma_0^{-2}, \quad \boldsymbol{q} = \boldsymbol{s} + \boldsymbol{z} \, ,$$

where $\theta_{x,y0}$ are defined by the vectors $\boldsymbol{n}_0$, $\boldsymbol{Z}_{\{hkl\}}$. The latter expressions prove that both multiple scattering and mosaicity make the same influence on radiation spectrum, which depends on the detector position. This influence, for example, is essentially suppressed for the photons emitted backwards the particle movement ($\theta_{\rm B} = \pi/2$).

In the kinematic approach, the PXR intensity from a single crystalline block is proportional to the crystallite thickness $L_i$, as follows from (2.50). Because the emission from separate crystalline blocks is incoherent, the spectral density of photons in the PXR reflection depends on the total sample thickness [17]:

$$\begin{aligned} \frac{\partial^3 N_{\{hkl\}}}{\partial\theta_{x0}\partial\theta_{y0}\partial\omega} &= \frac{\alpha}{2\pi}\frac{J}{e}\sum_{m=1}^{\infty}\int {\rm d}\boldsymbol{s}\,{\rm d}\boldsymbol{z}\,{\rm d}E\,\varphi_1\left(\frac{\boldsymbol{s}}{\Delta\theta_L}\right)\varphi_2\left(\frac{\boldsymbol{z}}{\nu}\right)\varphi_3\left(\frac{\epsilon}{\Delta E}\right) \\ &\times \frac{(\theta_{x0}^2 + q_x^2\cos^2\theta_{\rm B})\cos^2 2\theta_{\rm B} + (\theta_{y0}^2 + q_y^2\cos^2\theta_{\rm B})}{(\theta_{x0}^2 + \theta_{y0}^2 + q^2\cos^2\theta_{\rm B} + \gamma_0^{-2})^2}|\chi_{g_m}(\omega_{\rm B}^{(m)})|^2\frac{\omega_{\rm B}^{(m)} L}{c} \\ &\times\delta\left[\frac{2\sin^2\theta_{\rm B}}{\cos 2\theta_{\rm B}}\left(\omega - \omega_{\rm B}^{(m)} - \frac{\epsilon}{E_0}\gamma^{-2}\right) - (\theta_{x0} + q_x\cos\theta_{\rm B})\omega_{\rm B}^{(m)}\tan 2\theta_{\rm B}\right] . \end{aligned} \quad (2.51)$$

This result can be considered as an incoherent model for the formation of the PXR reflection in thick and mosaic crystals. We consider below the most important limiting case of (2.51), simplifying the equation in front of it. The electron beam in experiments is usually monochromatic enough and the energy loss for radiation in thin samples is small; hence the factor $\Delta E/E_0 \ll 1$ in (2.51) can be neglected. The inequality

$$\Delta\theta_L > \gamma_0^{-1} ,$$

quantifying the influence of multiple scattering on PXR, does not depend on the particle energy, as follows from (1.26):

$$\frac{E_{\mathrm{s}}}{E}\sqrt{\frac{L}{L_{\mathrm{R}}}} < \frac{m_e c^2}{E}, \qquad L < \frac{m_e c^2}{E_{\mathrm{s}}} L_{\mathrm{R}} \approx 6.25 \times 10^{-4} L_{\mathrm{R}} . \tag{2.52}$$

For majority of real crystals, the radiation length $L_{\mathrm{R}} \leq 1$ cm and inequality (2.52) is equivalent to the condition $L < L_{\mathrm{ext}}$ for the validity of kinematic approximation. Therefore, for the perfect mosaic sample, the angular distribution of the PXR intensity is mostly influenced by mosaicity and multiple scattering. These factors are well described by Gaussian distribution functions, which make it possible to integrate (2.51):

$$\frac{\partial^3 N_{\{hkl\}}}{\partial\theta_{x0}\partial\theta_{y0}\partial\omega} = \frac{\alpha}{2\pi}\frac{J}{e}\sum_{m=1}^{\infty}\int_{-\infty}^{\infty} \mathrm{d}q_y\, |\chi_{g_m}(\omega_{\mathrm{B}}^{(m)})|^2 \frac{L}{c\cos\theta_{\mathrm{B}}\tan 2\theta_{\mathrm{B}}}$$
$$\times \frac{(\theta_{x0}^2 + q_x^2\cos^2\theta_{\mathrm{B}})\cos^2 2\theta_{\mathrm{B}} + (\theta_{y0}^2 + q_y^2\cos^2\theta_{\mathrm{B}})}{(\theta_{x0}^2 + \theta_{y0}^2 + q^2\cos^2\theta_{\mathrm{B}})^2}\Phi\left(\frac{\boldsymbol{q}}{\nu_{\mathrm{eff}}}\right) ,$$

$$\Phi\left(\frac{\boldsymbol{q}}{\nu_{\mathrm{eff}}}\right) = \frac{1}{\pi\nu_{\mathrm{eff}}^2}\exp\left(-\frac{q_x^2+q_y^2}{\nu_{\mathrm{eff}}^2}\right), \qquad \nu_{\mathrm{eff}} = \sqrt{\nu^2+\theta_{\mathrm{s}}^2} ,$$
$$q_x = -\frac{\sin\theta_{\mathrm{B}}}{\omega_{\mathrm{B}}^{(m)}\cos^2\theta_{\mathrm{B}}}\left(\omega - \omega_{\mathrm{B}}^{(m)}\right) + \frac{\theta_{x0}}{\cos\theta_{\mathrm{B}}} . \tag{2.53}$$

This expression is a basis for evaluating PXR features in mosaic samples. Integrating (2.53) over the photon frequency, the formula for the diffraction peak, analogous to the one used in powder diffractometry [31], can be obtained:

$$\frac{\partial^2 N_{\{hkl\}}}{\partial\theta_{x0}\partial\theta_{y0}} = \frac{\alpha}{2\pi}\frac{J}{e}\frac{L\cos 2\theta_{\mathrm{B}}}{2\,c\,\sin^2\theta_{\mathrm{B}}}\Psi(\theta_{x0},\theta_{y0}) , \tag{2.54}$$

$$\Psi(\theta_{x0},\theta_{y0}) = \sum_{m=1}^{\infty}\int_{-\infty}^{\infty} \mathrm{d}q_x \mathrm{d}q_y\, \Phi\left(\frac{\boldsymbol{q}}{\nu_{\mathrm{eff}}}\right)|\chi_{g_m}(\omega_{\mathrm{B}}^{(m)})|^2$$
$$\times\, \omega_{\mathrm{B}}^{(m)}\frac{(\theta_{x0}^2 + q_x^2\cos^2\theta_{\mathrm{B}})\cos^2 2\theta_{\mathrm{B}} + (\theta_{y0}^2 + q_y^2\cos^2\theta_{\mathrm{B}})}{(\theta_{x0}^2 + \theta_{y0}^2 + q^2\cos^2\theta_{\mathrm{B}})^2} .$$

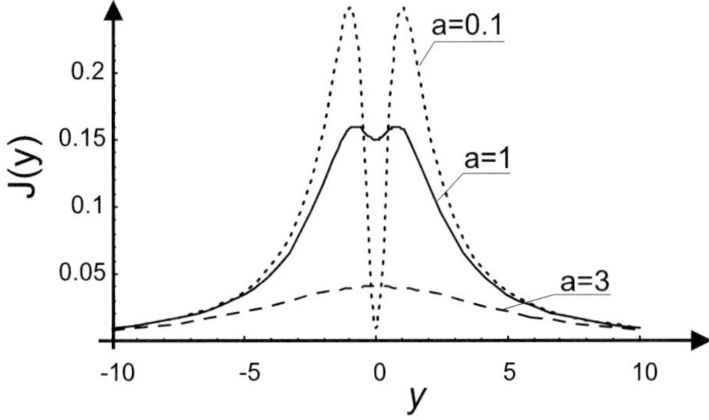


**Fig. 2.6.** Effect of the crystal mosaicity on the PXR angular distribution: $J(y) = \Psi(y\gamma^{-1}, 0)$, $\theta_B = 30°$, $a = \gamma^{-1}\sqrt{\nu^2 + \theta_s^2}$

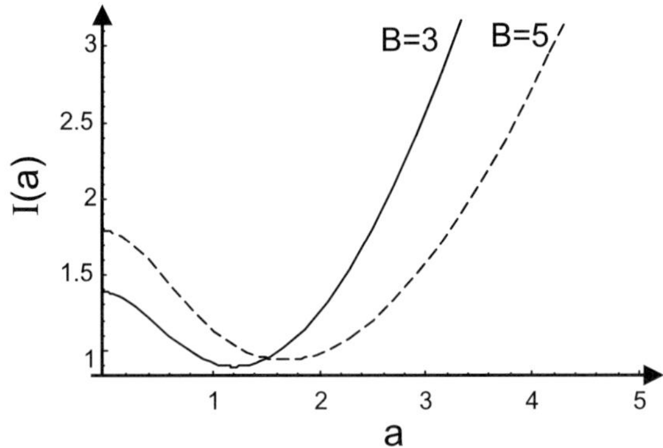


**Fig. 2.7.** Effect of the crystal mosaicity on the PXR intensity: $I(a)$ is the total intensity of the PXR reflection for different values of parameters $a$ and $B = \gamma^{-1}\theta_D$

The integrals in (2.54) are easily calculated numerically. Figure 2.6 demonstrates the influence of multiple scattering and sample mosaicity effect on the angular distribution of the PXR reflection. As expected, both factors lead to the broadening of the PXR peak. The total number of photons in the reflection, which is obtained by integrating (2.54) over observation angles, is changed inessentially in comparison with the ideal single crystal sample of the same thickness (Fig. 2.7).

## 2.4 Parametric $\gamma$-Radiation in Thin Mössbauer Crystals

Formula (2.32) for the PXR spectral–angular distribution is valid for arbitrary dependence of X-ray polarizability on the radiation frequency. Since the frequency of PXR is tunable, the PXR peak can be positioned into the absorption region of the atoms of a crystallographic unit cell by rotating a sample.

In this situation, the calculation of X-ray polarizability must be corrected for anomalous dispersion [31], i.e. scattering of radiation on the internal atomic shells. This resonant case is very illustrative for Mössbauer crystals, when parametric nature of radiation is caused by coherent scattering of the electromagnetic field of the charged particle on the resonant transitions of nuclei. The first investigation of this effect in the Vavilov–Cherenkov radiation from electrons in a homogeneous Mössbauer medium has been done in [30], and parametric $\gamma$-radiation (PGR) in Mössbauer crystals in [12].

In the frequency interval $|\omega - \omega_{\mathrm{r}}| \sim \Gamma$, where $\omega_{\mathrm{r}}$ and $\Gamma$ are the frequency and the width of resonant transition, respectively, the contribution of the Compton component into polarizability is negligible, in comparison to nuclear (resonant) scattering. In (2.32), the summation over reciprocal lattice vectors can be substituted by one term, for which the following condition is fulfilled:

$$\omega_{\mathrm{r}} \approx m\omega_{\{hkl\}} = m\frac{\pi c}{d_{\{hkl\}}\sin\theta_{\mathrm{B}}}, \qquad \boldsymbol{g}_{\mathrm{r}} = \frac{2}{c}\omega_{\mathrm{r}}\sin\theta_{\mathrm{B}}\boldsymbol{Z}_{\{hkl\}} . \tag{2.55}$$

In the frequency interval close to (2.55), the polarizability of a Mössbauer crystal depends mainly on the amplitude of coherent scattering of photons on nuclei. This amplitude can be found using general methods, developed in [1, 14], for the diffraction of resonant $\gamma$-radiation (see also Appendix A.1). Applying this method (2.32) for PGR in a thin crystal is modified to [17]

$$\begin{aligned} W_{\boldsymbol{g}}^{(s)} &= \frac{q^2}{4\pi^2\omega_{\mathrm{r}}^2 c^3}\left| n_{\boldsymbol{g}_{\mathrm{r}}}\frac{2K+1}{2K_0+1}Q(\boldsymbol{k},\boldsymbol{k}_g)\frac{(\boldsymbol{j}_k\boldsymbol{e}_{\mathrm{s}})([\boldsymbol{p}_g\boldsymbol{v}][\boldsymbol{p}_g\boldsymbol{j}_{k_g}])}{2(\omega-\omega_{\mathrm{r}}+\mathrm{i}\Gamma/2)}\right|^2 \\ &\times \frac{1}{[(\boldsymbol{k}-\boldsymbol{g})_\perp^2 + \frac{\omega^2}{v^2\gamma^2}]^2}\,\frac{4\sin^2(\omega/v - k_z + g_z)L/2}{v^2(\omega/v - k_z + g_z)^2} . \end{aligned} \tag{2.56}$$

Here $n_{\boldsymbol{g}_{\mathrm{r}}}$ is a Fourier component of the density distribution of resonant nuclei; $\boldsymbol{k}_g = \boldsymbol{k}+\boldsymbol{g}$; $K$ and $K_0$ are the moments of principal and excited nucleus states, respectively, and $\boldsymbol{j}_k$ is a matrix element of the electromagnetic current, which defines the resonant transition between principal $|0\rangle$ and excited $|1\rangle$ nucleus states:

$$\boldsymbol{j}_k = \langle 1|e(\hat{\boldsymbol{v}})_n \mathrm{e}^{-\mathrm{i}\boldsymbol{k}\boldsymbol{r}_n}|0\rangle ,$$

where $|\boldsymbol{j}_0|^2 = c\Gamma_1/2$ and $\Gamma_1$ is an elastic width of transition.

The Debye–Waller factor $Q(\boldsymbol{k},\boldsymbol{k}_g)$ determines the scattering probability of $\gamma$-quantum on the nucleus without recoil, i.e. without excitation of oscillations. The value of the factor depends on the characteristic frequency $\omega_{\mathrm{ph}}$ of phonon spectrum [1] and on the mean square $\overline{u^2}$ of the amplitude of nucleus oscillations:

$$Q(\boldsymbol{k},\boldsymbol{k}_g) \approx \mathrm{e}^{-1/2[W(\boldsymbol{k})+W(\boldsymbol{k}_g)]}, \qquad \Gamma < \omega_{\mathrm{ph}} ,$$

$$Q(\boldsymbol{k},\boldsymbol{k}_g) \approx \mathrm{e}^{-W(\boldsymbol{g})}, \qquad \Gamma > \omega_{\mathrm{ph}}, \qquad W(\boldsymbol{k}) = k^2\overline{u^2} .$$

Performing a summation in (2.56) over polarizations and introducing angular variables (2.39), the following expression is obtained for the PGR intensity from current $J$ transmitting through the Mössbauer crystal:

$$\frac{\partial^3 N_\gamma}{\partial\theta_x \partial\theta_y \partial\omega} = \frac{\alpha}{4\pi^2}\frac{Jc^2}{e}\left| n_{\boldsymbol{g}_\mathrm{r}} \frac{2K+1}{2K_0+1} Q(\boldsymbol{k},\boldsymbol{k}_g)\right|^2$$

$$\times \frac{[(\boldsymbol{tj}_k)(\boldsymbol{tj}_{k_g}) - t^2(\boldsymbol{j}_k^\perp \boldsymbol{j}_{k_g}^\perp)](\boldsymbol{j}_k \boldsymbol{j}_{k_g})}{4\omega_\mathrm{r}^5[(\omega-\omega_\mathrm{r})^2 + \Gamma^2/4][\theta_x^2+\theta_y^2+\gamma^{-2}]^2} \frac{4\sin^2(\omega/v - k_z + g_z)L/2}{(\omega/v - k_z + g_z)^2} . \quad (2.57)$$

Here, the vector $\boldsymbol{t}$ with components $(\theta_x \cos 2\theta_\mathrm{B}; \theta_y)$ and projection of the transmission current $\boldsymbol{j}_k^\perp$ are in the plane, which is perpendicular to the vector $(\omega_\mathrm{r}\boldsymbol{v}/v + \boldsymbol{g})$, pointing to the excited PXR reflection. To derive an angular distribution of resonant $\gamma$-quanta, (2.57) has to be integrated over the frequencies. In the interval $\Delta\omega \geq \Gamma$, the kinematic factor $\sin^2(\omega/v - k_z + g_z)L/2$ is assumed to be constant because the condition $\Gamma L < \Gamma L_\mathrm{ext} \ll 1$ is valid for real crystals. Thus, the PGR intensity is

$$\frac{\partial^2 N_\gamma}{\partial\theta_x \partial\theta_y} = \frac{\alpha}{2\pi}\frac{Jc^2}{e}\left| n_{\boldsymbol{g}_\mathrm{r}} \frac{2K+1}{2K_0+1} Q(\boldsymbol{k},\boldsymbol{k}_g)\right|^2$$

$$\times \frac{[(\boldsymbol{tj}_k)(\boldsymbol{tj}_{k_g}) - t^2(\boldsymbol{j}_k^\perp \boldsymbol{j}_{k_g}^\perp)](\boldsymbol{j}_k \boldsymbol{j}_{k_g})}{\omega_\mathrm{r}^4 \Gamma[\theta_x^2+\theta_y^2+\gamma^{-2}]^2} \frac{\sin^2 u_\mathrm{r} L/2}{u_\mathrm{r}^2} ;$$

$$u_\mathrm{r} = \frac{2\sin^2\theta_\mathrm{B}}{c\cos 2\theta_\mathrm{B}}(\omega_\mathrm{r} - \omega_\mathrm{B}) + \frac{\omega_\mathrm{r}}{c}\theta_x \tan 2\theta_\mathrm{B} . \quad (2.58)$$

The maximum of the PGR intensity is emitted in the direction

$$\theta_y = 0, \qquad \theta_x = \tan\theta_\mathrm{B}(\omega_\mathrm{r} - \omega_\mathrm{B}) ,$$

and the peak has an angular width

$$\Delta\theta_y \simeq \gamma^{-1}; \qquad \Delta\theta_x \simeq \frac{2c}{\omega_\mathrm{r} L \tan 2\theta_\mathrm{B}} . \quad (2.59)$$

The frequency dependence of PGR is determined by the width of resonant transition. However, the peak amplitude depends on the proximity of the Bragg frequency $\omega_\mathrm{B}$ to $\omega_\mathrm{r}$:

$$\frac{\partial N_\gamma}{\partial\omega} = \frac{\alpha}{2\pi}\frac{Jc^3}{e} L \left| n_{\boldsymbol{g}_\mathrm{r}} \frac{2K+1}{2K_0+1} Q(\boldsymbol{k},\boldsymbol{k}_g)\right|^2 \frac{\cos 2\theta_\mathrm{B}}{2\sin^2\theta_\mathrm{B}}$$

$$\times \frac{[(\boldsymbol{tj}_k)(\boldsymbol{tj}_{k_g}) - t^2(\boldsymbol{j}_k^\perp \boldsymbol{j}_{k_g}^\perp)](\boldsymbol{j}_k \boldsymbol{j}_{k_g})}{\omega_\mathrm{r}^6[(\omega-\omega_\mathrm{r})^2 + \Gamma^2/4]\left[\frac{\cos^2 2\theta_\mathrm{B}}{4\sin^4\theta_\mathrm{B}}(1-\omega_\mathrm{B}/\omega_\mathrm{r})^2 + \gamma^{-2}\right]^{1/2}} . \quad (2.60)$$

Thus, PGR allows us to produce a very convergent resonant radiation that is necessary in many applications of Mössbauer radiation (for example, [3]).

**Table 2.2.** Parameters for Mössbauer crystals ($\chi_0 = -\mu/2\frac{\Gamma}{(\omega-\omega_{\rm r}+{\rm i}\Gamma/2)}$)

| Crystal | $\hbar\omega_{\rm r}$ (keV) | $\Gamma/\omega_{\rm r}$ | $\mu$ | $L_{\rm abs}$ (cm) | $(hkl)$ | $2\theta_{\rm B}$ (degree) |
|---|---|---|---|---|---|---|
| | | | | | (0 0 2) | 27.2 |
| $Fe^{57}$ | 14.4 | $3\times10^{-13}$ | $2.6\times10^{-4}$ | $5\times10^{-5}$ | (1 1 1) | 23.5 |
| | | | | | (8 2 2) | 178 |
| | | | | | (0 1 1) | 6.8 |
| $W^{183}$ | 47.2 | $6.5\times10^{-11}$ | $2.9\times10^{-6}$ | $1.4\times10^{-4}$ | (0 0 2) | 9.7 |
| | | | | | (21 11 0) | 177 |

As follows from (2.59), the divergence of the PGR beam from the crystal of thickness $L = 10^{-4}$ cm and Mössbauer nuclei $Fe^{57}$ ($\omega_{\rm r} = 14.4\,$keV) is given by

$$\Delta\theta_x \simeq 2\times10^{-5}\ \text{rad}\ .$$

The maximal number of resonant $\gamma$-quanta in the PGR peak is estimated by integrating (2.58) over the angles under the condition $\omega_{\rm B} = \omega_{\rm r}$:

$$N_\gamma \simeq \frac{\alpha}{2\omega_{\rm r}^6}\frac{Jc^5}{e}L\gamma\left|n_{g_{\rm r}}\frac{2K+1}{2K_0+1}Q(\boldsymbol{k},\boldsymbol{k}_g)\right|^2\frac{\cos^2 2\theta_{\rm B}}{\sin 2\theta_{\rm B}\sin^2\theta_{\rm B}}\frac{\Gamma_1^2}{\Gamma}\ . \tag{2.61}$$

Table 2.2 contains all necessary parameters of Mössbauer iron $Fe^{57}$ and tungsten $W^{183}$ crystals, which are required for simulation of PGR spectra. For instance, the total number of photons (2.61) in the PGR peak from the resonant transition $Fe^{57}(\Gamma/\omega_{\rm r} \simeq 3\times10^{-13}, \Gamma_1/\Gamma \simeq 0.2)$ [30] with $L = 10^{-4}$ cm and the angular width (2.59) corresponds to the source with the brightness

$$I_{\rm PGR} \approx 3.4\times10^{17} J[{\rm A}]\ E[{\rm GeV}]\frac{\text{photons}}{\text{s mrad}^2(0.1\%\ \text{bandwidth})}\ .$$

The PXR in the vicinity of the atomic absorption edge has been recently considered in [29].

## References

1. A.M. Afanas'ev, Yu.M. Kagan: Zh. Eksp. Teor. Fiz. **48**, 327 (1965)
2. A.I. Akhiezer, N.F. Shul'ga: Usp. Fiz. Nauk **137**, 561 (1982)
3. M.A. Andreeva, V.G. Semenov, B. Lindgren, L. Häggström, B. Kalska, A.I. Chumakov, O. Leupold, R. Rüffer, K.A. Prokhorov, N.N. Salashchenko: Hyperfine Interact. **141/142**, 119 (2002)
4. A. Authier: *Dynamical Theory of X-ray Diffraction* (Oxford University Press, New York 2001) pp 3–26
5. V.N. Baier, V.M. Katkov, V.S. Fadin: *Radiation from Relativistic Electrons* (Atomizdat, Moscow 1973)

6. V.N. Baier, V.M. Katkov, V.M. Strakhovenko: *Electromagnetic Processes at High Energies in Oriented Single Crystals* (World Scientific, Singapore 1998)
7. V.G. Baryshevsky: Dokl. Akad. Sci. Belarus **15**, 306 (1971); V.G. Baryshevsky, I.D. Feranchuk: *Proc. XXI Conf. on Nuclear Structure and Nuclear Spectroscopy* (Moscow University, Moscow 1971) p 220
8. V.G. Baryshevsky: *Channelling, Radiation and Reactions at Crystals under High Energy* (Belarussian State University, Minsk 1982)
9. V.G. Baryshevsky, V.A. Danilov, O.L. Ermakovich, I.D. Feranchuk, A.V. Ivashin, V.I. Kozus, S.G. Vinogradov: Phys. Lett A **110**, 477 (1985)
10. V.G. Baryshevsky, I.D. Feranchuk: Zh. Eksp. Teor. Fiz. **61**, 944 (1971) (Sov. Phys. JETP **34**, 502 (1972)); Erratum: ibid **64**, 760 (1973) (Sov. Phys. JETP **36**, 399 (1973))
11. V.G. Baryshevsky, I.D. Feranchuk: Dokl. Belarusian Acad. Sci. **18**, 499 (1974)
12. V.G. Baryshevsky, I.D. Feranchuk: J. Phys. (Paris) **44**, 913 (1983)
13. V.G. Baryshevsky, I.D. Feranchuk, A.O. Grubich, A.V. Ivashin: Nucl. Instrum. Methods A **249**, 306 (1986)
14. B. Batterman, H. Cole: Rev. Mod. Phys. **36**, 681 (1964)
15. B.A. Bazylev, N.K. Zhevago: *Radiation from High Energy Particles in Media and External Fields* (Nauka, Moscow 1988)
16. M. Born, E. Wolf: *Principles of Optics* (Pergamon Elmsford, New York 1975) p 328
17. I.D. Feranchuk: Coherent phenomena in the processes of X-ray and gamma-radiation from relativistic charged particles in crystals. Habilitation Doctor Thesis, Belarussian State University, Minsk (1984)
18. I.D. Feranchuk, A.V. Ivashin: J. Phys. (Paris) **46**, 1981 (1985)
19. G.M. Garibyan, C. Yang: *X-ray Transition Radiation* (Armenian Academy of Sciencies, Erevan 1983)
20. V.L. Ginzburg: *Theoretical Physics and Astrophysics* (Nauka, Moscow 1975) pp 120–160
21. V.L. Ginzburg, I.M. Frank: Zh. Eksp. Teor. Fiz. **16**, 15 (1946)
22. R.W. James: *The Optical Principles of the Diffraction of X-rays* (G. Bell and Sons, London 1950)
23. N.P. Kalashnikov: *Coherent Interaction of Charged Particles in Monocrystals* (Atomizdat, Moscow 1981)
24. J. Kirz, D.T. Attwood, B.L. Henke: *X-ray Data Booklet* (Lawrence Berkeley Laboratory, Berkeley 2001)
25. M.A. Kumakhov: *Radiation from the Channelled Particle in Crystals* (Energoatomizdat, Moscow 1986)
26. L.D. Landau, E.M. Lifshitz: *Classical Theory of Fields* (Pergamon, New York 1959)
27. L.D. Landau, E.M. Lifshitz: *Electrodynamics of Continuous Media* (Nauka, Moscow 1982)
28. P. Morse, H. Feshbach: *Methods of Theoretical Physics* (McGraw-Hill, New York 1953) p 735
29. H. Nitta: Nucl. Instrum. Methods B **115**, 401 (1996)
30. E.A. Perelshtein, M.I. Podgoretzky: Yad. Fiz. **12**, 1149 (1970)
31. M.A. Porai-Koshitz: *Practical Course for X-ray Structure Analysis, vol 2* (Fizmatgiz, Moscow 1968)
32. V.P. Silin, A.A. Rukhadze: *Electromagnetic Properties of Plazma* (Gosatomizdat, Moscow 1961)

33. M.L. Ter-Mikaelian: *High Energy Electromagnetic Processes in Condensed Media* (in Russian: AN ArmSSR, Yerevan 1969) (in English: Wiley, New York 1972)
34. D.F. Walls, G.J. Milburn: *Quantum Optics* (Springer, Berlin 1994)

# 3

# Dynamical Theory of Parametric X-ray Radiation

## 3.1 Quantum Electrodynamics for Radiation Processes in Crystals

Previous chapters demonstrate a close analogue between the theory of parametric X-ray radiation (PXR) in thin crystals and the kinematic theory of X-ray diffraction, which is used in X-ray analysis, viz. powder diffractometry, phase identification, contamination analysis, X-ray topography, etc. (see, for example, [12]). Complementary to kinematic domain, a high-resolution X-ray diffraction (HRXRD), based on the dynamical diffraction theory, opens a new horizon for investigation of fine structures of modern materials and nanoscale objects [2, 16]. By analogy, PXR in thick crystals is expected to extend its applicability to dynamical domain, too. The crystal is considered to be thick if the thickness of the sample $L \gg L_{\text{ext}}$, i.e. for the X-ray range $L \geq 1\,\mu\text{m}$.

The dynamical theory of PXR [4, 5, 9, 14, 15] is constructed on the basis of special representation of quantum electrodynamics (QED), developed for radiation processes in crystals. Generally, the quantum effects are not needed to be taken into account in simulation of X-ray radiation from charged particles due to negligible recoil effects $\hbar\omega \ll E$. However, the use of QED for the PXR theory has at least two advantages: (i) the quantum approach at $\hbar\omega \ll E$ is still simple as the classic one, but the results are valid for any energy; (ii) a clear and straightforward classification of all radiation processes in crystals and their relation to PXR can be built.

The quantum description of the electromagnetic processes in a crystal is based on the following representation of the Hamiltonian of the system:

$$\hat{H} = \hat{H}_{e,c} + \hat{H}_{\gamma,c} + \hat{H}_{e,\gamma}\ , \tag{3.1}$$

which differs from the vacuum QED [1] because the Hamilton operators for the electron and photon subsystems include both the operators $\hat{H}_e, \hat{H}_\gamma$ of the free fields and the coherent potentials of their interaction with a crystal. Therefore, the second quantization of these fields is based not on the vacuum plane waves but on the one-particle eigenfunctions of the operators $\hat{H}_{e,c}$, $\hat{H}_{\gamma,c}$:

$$\hat{\Psi}(\boldsymbol{r}) = \sum_{\nu} [b_\nu \Psi^{+}_{\nu E}(\boldsymbol{r}) + b^{+}_{\nu} \Psi^{-}_{\nu E}] ;$$
$$\hat{\boldsymbol{A}}(\boldsymbol{r}) = \sum_{\mu} [a_\mu \boldsymbol{A}^{+}_{\mu\omega}(\boldsymbol{r}) + a^{+}_{\mu} \boldsymbol{A}^{-}_{\mu\omega}] . \quad (3.2)$$

Here $b_\nu$ ($b^{+}_{\nu}$) and $a_\mu$ ($a^{+}_{\mu}$) are the annihilation (creation) operators of electrons and photons, respectively. Both particles are in stationary states, which are described by the wavefunction $\Psi^{-}_{\nu E}(\Psi^{+}_{\nu E})$ or by the vector potential $\boldsymbol{A}^{-}_{\mu\omega}(\boldsymbol{A}^{+}_{\mu\omega})$ for electrons and photons, respectively. The quantization procedure for the electromagnetic field in absorbing crystals is presented in detail in [4, 9]. For PXR, the functions $\Psi^{-}_{\nu E}$ and $\boldsymbol{A}^{-}_{\mu\omega}$ for the particles in final states have to be found as solutions for boundary conditions with convergent spherical waves. The functions $\Psi^{+}_{\nu E}$ and $\boldsymbol{A}^{+}_{\mu\omega}$ for initial states correspond to the solutions with divergent waves. This fact is the generalization of the well known reciprocity theorem, which is applied for radiation and scattering processes of electromagnetic waves [7].

Below we derive explicit functions required for simulation of PXR from an electron in a crystal. The wavefunctions $\Psi^{\pm}_{E}(\boldsymbol{r})$ for the electron states with the energy $E$ and quantum number $\nu$ are defined as the solutions of the Dirac equation:

$$\{(\alpha_l \hat{p}_l)c + \beta mc^2 + U(\boldsymbol{r}) - E\}\Psi^{\pm}_{\nu E}(\boldsymbol{r}) = 0 . \quad (3.3)$$

Here $\alpha_l$, $l = 1, 2, 3$, $\beta$ are the Dirac matrices; $\hat{p}_l = -\mathrm{i}\hbar\partial/\partial x_l$ is the momentum operator. The potential $U(\boldsymbol{r})$ describes the coherent interaction of the electron with the periodical field of a crystal and is expressed through the amplitudes of its elastic scattering by the atoms of the crystallographic unit cell [17]:

$$U(\boldsymbol{r}) = \sum_{\boldsymbol{g}} U_{\boldsymbol{g}} \mathrm{e}^{\mathrm{i}\boldsymbol{g}\boldsymbol{r}} H(z)H(L-z) ;$$
$$U_{\boldsymbol{g}} = -\frac{4\pi e^2}{g^2 \Omega} \sum_{j} [Z_j - F_j(\boldsymbol{g})] \, \mathrm{e}^{\mathrm{i}\boldsymbol{g}\boldsymbol{R}_j} \mathrm{e}^{-W_j(\boldsymbol{g})} , \quad (3.4)$$

where $\boldsymbol{g}$ is the reciprocal lattice vector; $\boldsymbol{R}_j$ is the coordinate of the atom with index $j$ in the cell; $Z_j$, and $F_j(\boldsymbol{g})$ are the nucleus charge and atomic scattering factor, respectively; $\mathrm{e}^{-W_j(\boldsymbol{g})}$ is the Debye–Waller factor and $\Omega$ is the volume of the unit cell.

The full set of the states for QED in the media is constructed for the crystal of finite thickness $L$, which is taken into account in (3.4) by means of the Heaviside functions $H(z)$ and $H(L-z)$.

On the other hand, the stationary states of the electromagnetic field with the frequency $\omega$ and quantum numbers $\mu$ in the crystal are defined by the vector potentials $\boldsymbol{A}^{\pm}_{\mu\omega}(\boldsymbol{r})$, which are the solutions of Maxwell's equations with the periodic dielectric permittivity of the crystal [2]:

$$\epsilon_{\rm ij}(\boldsymbol{r},\omega) = \delta_{\rm ij} + \chi_{\rm ij}(\boldsymbol{r},\omega)H(z)H(L-z)\ . \qquad (3.5)$$

The formulas for calculation of the X-ray polarizability $\chi_{\rm ij}(\boldsymbol{r},\omega)$ are derived in Appendix A.1:

$$\chi_{\rm ij}(\boldsymbol{r},\omega) = \delta_{\rm ij}\sum_{\boldsymbol{g}} \chi_{\boldsymbol{g}} {\rm e}^{{\rm i}\boldsymbol{g}\boldsymbol{r}}\ ;$$

$$\chi_{\boldsymbol{g}} = -\frac{4\pi e^2}{m\omega^2\Omega}\sum_j [F_j(\boldsymbol{g}) + \ f'(\omega) + {\rm i}f''(\omega)]{\rm e}^{{\rm i}\boldsymbol{g}\boldsymbol{R}_{\rm j}}{\rm e}^{-W_j(\boldsymbol{g})}\ , \qquad (3.6)$$

where $f'$ and $f''$ are the anomalous dispersion corrections.

The Hamiltonian $\hat{H}_{e,\gamma}$ for the interaction between the electron and electromagnetic fields in a crystal has the same form as in the vacuum QED [1]:

$$\hat{H}_{e,\gamma} = e\int {\rm d}\boldsymbol{r}\ \hat{\Psi}^*(\boldsymbol{r})(\alpha_l \hat{A}_l(\boldsymbol{r}))\hat{\Psi}(\boldsymbol{r})\ , \qquad (3.7)$$

with the field operators (3.2). Similar to the vacuum QED, the amplitude of any electromagnetic process in the crystal, homogenous media or in external fields is calculated using a perturbation theory over the operator $\hat{H}_{e,\gamma}$. This results in a similar to vacuum diagrams for amplitudes, however, with modified physical meaning of lines. Instead of thin lines corresponding to the plane waves in the vacuum QED, the thick lines, defining the wavefunctions of the particle or the photon and taking into account their interaction with the media, have to be used. Each diagram uses the same vertex as in the case of vacuum QED. If multiple photon processes are neglected, the Feynman diagrams of the first order give the most essential contribution to the radiation yield. In Fig. 3.1, diagram (a) corresponds to the process in the vacuum QED, which is evidently forbidden due to conservation of momentum and energy [1]. Other diagrams describe the following radiation processes: (b) Cherenkov radiation in homogeneous media with the constant refraction index $n$; (c) synchrotron radiation from the electron in the magnetic field $\boldsymbol{H}$; (d) PXR; (e) coherent bremsstrahlung and radiation from channelled particles; and (f) a general case of the radiation in a crystal.

The reconstruction of an analytical expression for the scattering amplitude from the diagram is also similar to the vacuum QED, for example, for diagram (f) in Fig. 3.1 [5]:

$$M_{fi} = 2\pi\delta(E - E_1 - \omega)T_{fi};\quad T_{fi} = e\int {\rm d}\boldsymbol{r}\ (\Psi^-_{\nu_1 E_1})^*(\alpha_l A^-_{l\mu\omega})^*(\Psi^+_{\nu E}).\ (3.8)$$

## 3.2 Analytical Expressions for the Electron Wavefunction and the Vector Potential of the X-ray Field in a Crystal

In order to calculate the matrix elements (3.8), the expressions for electron and photon eigenstates have to be derived, and several approximations are used

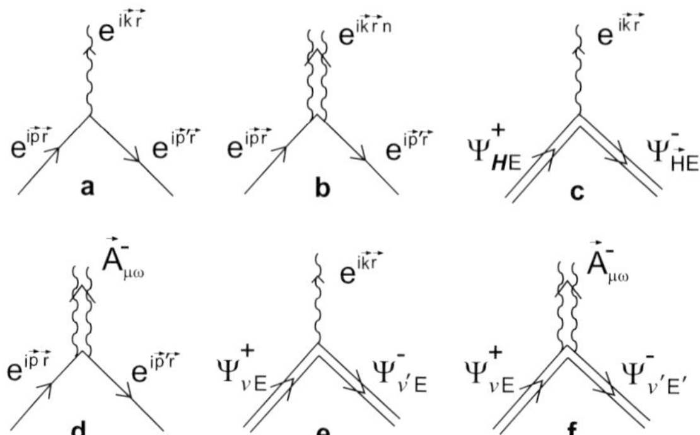


**Fig. 3.1.** Feynman diagrams for the quantum electrodynamics in media

here for this derivation. Moreover, the solutions for homogeneous equations in the quantum case are found similarly to the classic approach on the basis of (2.16).

The electron wavefunction can be found as a solution of (3.3). A typical value of the crystal potential $|U(\boldsymbol{r})| \leq 10$ eV and the electron energy $E \geq 100\,\text{keV}$, i.e.

$$|U(\boldsymbol{r})| \ll E\,, \tag{3.9}$$

which permits us to exclude [1] most of the spinor operators in (3.3) with the accuracy $O(U^2/E^2)$:

$$\left[-\hbar^2 c^2 \Delta + m^2 c^4 - E^2 + 2EU(\boldsymbol{r}) - \mathrm{i}c\hbar\left(\alpha_l \frac{\partial}{\partial x_l}\right) U(\boldsymbol{r})\right] \Psi_E(\boldsymbol{r}) = 0. \tag{3.10}$$

The stationary states in the infinite crystal are calculated using the perturbation theory (PT), which is applicable due to (3.9):

$$\begin{aligned} \Psi_E(\boldsymbol{r}) &\equiv \Psi_0 + \sum_{\boldsymbol{g}\neq 0} \Psi_{\boldsymbol{g}} \approx \mathrm{e}^{\mathrm{i}\boldsymbol{pr}} \left[1 + \sum_{\boldsymbol{g}\neq 0} \hat{B}_{\boldsymbol{g}} \mathrm{e}^{\mathrm{i}\boldsymbol{gr}}\right] \chi_E\,; \\ \hat{B}_{\boldsymbol{g}} &= -\frac{2E + (\alpha_l h_l)}{2(\boldsymbol{pg}) + g^2} U_{\boldsymbol{g}}; \qquad p_z = \sqrt{E^2 - m^2 - p_\perp^2}\,, \end{aligned} \tag{3.11}$$

where $\chi_E$ is the bispinor for the free electron [1]. The PT fails for only one special geometry when the particle momentum is perpendicular to the reciprocal lattice vector, i.e. $(\boldsymbol{pg}) \approx 0$. This case corresponds to the channelling of the electron, and therefore a nonperturbative method has to be used for the solution of (3.10) in order to take into account a zone spectrum of the transverse movement [4]. However, the case of channelling is not essential for PXR, and we assume $(\boldsymbol{pg}) \neq 0$ and $2(\boldsymbol{pg}) \gg g^2$ for relativistic electrons.

The continuity conditions for the wavefunction and its derivative, being applied at the crystal boundaries $z = 0, L$, result in

$$\Psi_E^{(\pm)} \approx \mathrm{e}^{\mathrm{i}\boldsymbol{pr}} \left[1 + \varphi_E^{(\pm)}(\boldsymbol{r})\right] \chi_E \, ;$$

$$\varphi_E^{(+)} = \sum_{\boldsymbol{g} \neq 0} \hat{B}_{\boldsymbol{g}} \mathrm{e}^{\mathrm{i}\boldsymbol{gr}} \{C_g(z) H(z) H(L-z) + C_g(L) H(z-L)\} \, ;$$

$$C_g(z) = 1 - \mathrm{e}^{\mathrm{i}(\tilde{p}_g - p_z - g_z)z}; \qquad \tilde{p}_g = \sqrt{E^2 - m^2 - (\boldsymbol{p}_\perp - \boldsymbol{g}_\perp)^2} \, ;$$

$$\varphi_E^{(-)} = \sum_{\boldsymbol{g} \neq 0} \hat{B}_{-\boldsymbol{g}} \mathrm{e}^{-\mathrm{i}\boldsymbol{gr}} \{C_{-g}(z-L) H(z) H(L-z) + C_{-g}(-L) H(-z)\} \, . \quad (3.12)$$

The function $\Psi_E^+$ contains the incident plane wave on the left sample side, and $\Psi_E^-$ corresponds to the incident wave from the right side. Both waves are related [4, 9] by the expression

$$\Psi_E^-(\boldsymbol{p}, \boldsymbol{r}) = [\Psi_E^+(-\boldsymbol{p}, \boldsymbol{r})]^* \, . \quad (3.13)$$

The vector potential $\boldsymbol{A}_{\mu\omega}^{\pm}(\boldsymbol{r})$ is found from the diffraction of the emitted photon and the regular boundary conditions for the field at both surfaces of the crystal. For the potential $\boldsymbol{A}_{\mu\omega}^{+}(\boldsymbol{r})$, this procedure is equivalent to the solution of the conventional X-ray diffraction problem [2]. The function $\boldsymbol{A}_{\mu\omega}^{-}(\boldsymbol{r})$ is related to the former by the expression

$$\boldsymbol{A}_{\boldsymbol{k}\omega}^{-}(\boldsymbol{r}) = [\boldsymbol{A}_{-\boldsymbol{k}\omega}^{+}(\boldsymbol{r})]^*,$$

which is equivalent to the reciprocity theorem. Using the dynamical diffraction theory [2], the potential $\boldsymbol{A}_{\mu\omega}^{+}(\boldsymbol{r})$ for the Bragg geometry in a two-beam approximation is expressed as follows:

$$\begin{aligned} \boldsymbol{A}_{\boldsymbol{k}\omega}^{(+)s}(\boldsymbol{r}) \equiv \boldsymbol{A}_{\boldsymbol{k}} + \boldsymbol{A}_{\boldsymbol{g}} = \sqrt{4\pi} \mathrm{e}^{\mathrm{i}\boldsymbol{kr}} \big\{ & (\boldsymbol{e}_{\mathrm{s}} + \boldsymbol{e}_{sg} \mathrm{e}^{\mathrm{i}\boldsymbol{gr}} D_{sg}(L)) H(-z) \\ & + (\boldsymbol{e}_{\mathrm{s}} D_{s0}(L-z) + \boldsymbol{e}_{sg} \mathrm{e}^{\mathrm{i}\boldsymbol{gr}} D_{sg}(L-z)) H(z) H(L-z) \\ & + \boldsymbol{e}_{\mathrm{s}} D_{s0}(L) H(z-L) \big\} \, ; \end{aligned}$$

$$D_{s0}(z) = -\sum_{\mu=1}^{2} \gamma_{\mu s}^{0} \mathrm{e}^{\mathrm{i}\omega\epsilon_{\mu s} L \gamma_0^{-1}}; \qquad D_{sg}(z) = \beta \sum_{\mu=1}^{2} \gamma_{\mu s}^{g} \mathrm{e}^{\mathrm{i}\omega\epsilon_{\mu s} L \gamma_0^{-1}} \, . \quad (3.14)$$

Here $\boldsymbol{A}_{\boldsymbol{k}}$ and $\boldsymbol{A}_{\boldsymbol{g}}$ are the primary and diffracted wave fields, respectively (Fig. 3.2), and the conventional notation of the dynamical diffraction theory [2] is used: $\boldsymbol{e}_{\mathrm{s}}, \boldsymbol{e}_{sg}$; $s = 1, 2$ are the polarization vectors for the primary and diffraction waves; $\gamma_0 = (\boldsymbol{kN})/k$, where $\boldsymbol{N}$ is the normal to the crystal surface; $\beta = \gamma_0/\gamma_g$; $\gamma_g = (\boldsymbol{k}_g \boldsymbol{N})/k_g$; $\boldsymbol{k}_g = \boldsymbol{k} + \boldsymbol{g}$. The expressions for the vector potential for the different PXR geometries are compiled in Appendix A.2.

The solutions of the dispersion equation, which defines the refraction of the waves in a crystal, are expressed [2] through X-ray polarizability (3.6):

$$\epsilon_{\mu\mathrm{s}} = \frac{1}{4} \{ -\alpha_{\mathrm{B}} + \chi_0(\beta + 1) \pm \sqrt{[-\alpha_{\mathrm{B}} + \chi_0(\beta - 1)]^2 + 4\beta C_{\mathrm{s}}^2 \chi_h \chi_{-h}} \} \, ;$$

$$C_{\mathrm{s}} = \cos 2\theta_{\mathrm{B}}; \qquad \sin\theta_{\mathrm{B}} = \frac{g}{2k}; \qquad \alpha_{\mathrm{B}} = \frac{2\boldsymbol{kg} + g^2}{k^2}. \quad (3.15)$$

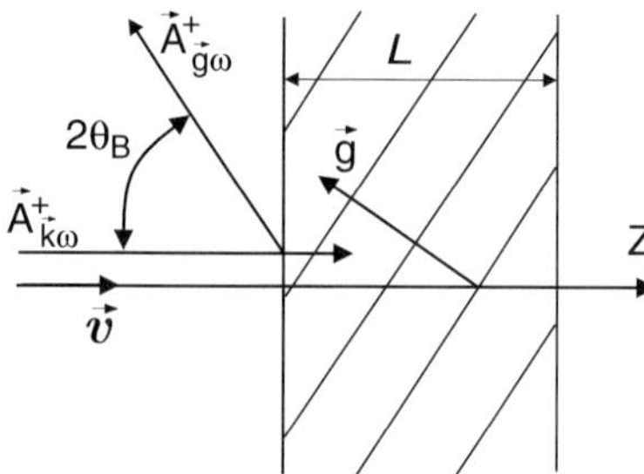


**Fig. 3.2.** Wave propagation for the diffraction in the Bragg geometry

The dimensionless parameter $\alpha_{\mathrm{B}}$ is a deviation of the wave vector of the emitted photon from the exact Bragg condition. The parameters for a linear combination of exponents in (3.14) are

$$\gamma^0_{1(2)s} = \frac{2\epsilon_{2(1)s} - \chi_0}{2(\epsilon_{(2(1)s} - \chi_0) - 2(\epsilon_{1(2)s} - \chi_0)\exp[\mathrm{i}\omega(\epsilon_{2(1)s} - \epsilon_{1(2)s})L/\gamma_0]} ;$$

$$\gamma^g_{1(2)s} = \frac{-\beta C_{\mathrm{s}}\chi_h}{2(\epsilon_{(2(1)s} - \chi_0) - 2(\epsilon_{1(2)s} - \chi_0)\exp[\mathrm{i}\omega(\epsilon_{2(1)s} - \epsilon_{1(2)s})L/\gamma_0]} . \quad (3.16)$$

Thus, (3.11)–(3.16) determine the eigenfunctions inside and outside the crystal and are used for calculation of the matrix element (3.8).

## 3.3 Calculation of the Parametric X-ray Radiation and Coherent Bremsstrahlung Intensities

In general, both PXR and CBS behave as shown in Fig. 3.1f. The differential cross-section of the radiation process in Fig. 3.1f with an initial electron of energy $E$ and momentum $\boldsymbol{p}$ and the electron $(\boldsymbol{p}_1, E_1)$ and the photon $(\boldsymbol{k}, \omega)$ in the final state follows from (3.8) using a standard QED technique [1,4]:

$$\mathrm{d}\sigma = 2\pi\delta(E - E_1 - \omega)|T_{fi}|^2 \frac{\mathrm{d}\boldsymbol{k}\mathrm{d}\boldsymbol{p}_1}{8\hbar\omega(2\pi)^6 pE_1} . \quad (3.17)$$

The number of photons emitted during the unit time period, spectral interval $\mathrm{d}\omega$ and solid angle $\mathrm{d}\Omega$ in the vicinity of the vector $\boldsymbol{n} = c\boldsymbol{k}/\omega$ from the electron beam with the current $J$ and cross-section $S$ is

$$\frac{\partial^2 N}{\partial\omega\partial\boldsymbol{n}} = \frac{J\omega}{eS\hbar c^2}\sum\int |T_{fi}|^2\delta(E - E_1 - \omega)\frac{\mathrm{d}\boldsymbol{p}_1}{8(2\pi)^5 pE_1} , \quad (3.18)$$

where the integration over $E_1, \boldsymbol{p}_1$, averaging over the electron spin in the initial state, and summation over the final states have been performed.

As mentioned in Chap. 1, the angular distribution of radiation from a relativistic electron is concentrated within the narrow cones with the angular width $\Delta\theta \sim mc^2/E$, one along the electron velocity and others along the

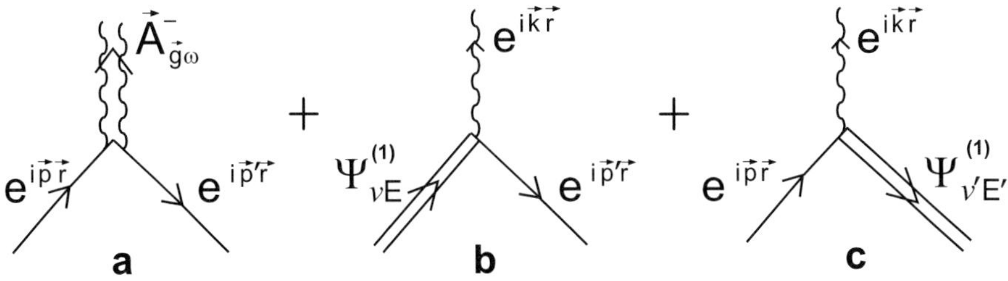


**Fig. 3.3.** Diagrams for the CBS+PXR radiation

reciprocal lattice vectors. Each of these PXR reflections possesses its own wavefunctions, corresponding to the vector $\boldsymbol{g}$. Figure 3.3 shows the diagrams contributing to PXR side reflections: (a) is the diffraction of emitted photons into the PXR peak, (b) and (c) are the CBS from the diffracted electrons in the initial and final states, respectively. The picture demonstrates the difference between the processes: Despite the large momentum transfer during the production of photons with the wave vector $(\boldsymbol{k}+\boldsymbol{g})$ in all cases, the momentum recoil for PXR is received by the crystal and for CBS it is compensated by the change of the electron momentum. This compensation is due to electron diffraction on the crystallographic planes with the reciprocal lattice vector $-\boldsymbol{g}$. The electron diffraction peaks, however, are not resolved because they are allocated within the narrow cone along the electron velocity vector due to the relativistic energy of electrons and the condition $p \gg \hbar g$. Nevertheless, the secondary effects such as CBS are seen because of the large scattering angle for the emitted photons due to $E \gg \hbar\omega$. Using the identity

$$\left| \int \mathrm{d}\boldsymbol{r}_\perp \mathrm{e}^{\mathrm{i}(\boldsymbol{p}-\boldsymbol{p}_1-k)\boldsymbol{r}_\perp} \right|^2 = 2\pi S \delta(\boldsymbol{p}_\perp - \boldsymbol{p}_{1\perp} - k_\perp) ,$$

equation (3.18) is transformed to the following:

$$\frac{\partial^2 N_{gs}}{\partial\omega \partial\, \boldsymbol{n}} = \frac{\alpha\omega}{4\pi^2 c^4} (\boldsymbol{e}_{gs}\boldsymbol{v}_g)^2 \frac{J}{e} |M_{\mathrm{PXR}} + M_{\mathrm{CBS}}|^2 ;$$

$$M_{\mathrm{PXR}} = \sum_{\mu=1}^{2} \gamma_{\mu s}^{g} \left( \frac{1}{q_{0g}} - \frac{1}{q_{\mu g s}} \right) \left( \mathrm{e}^{\mathrm{i} q_{\mu g s} L/\gamma_1} - 1 \right) ;$$

$$M_{\mathrm{CBS}} = -\frac{U_{-\boldsymbol{g}}}{\hbar(\boldsymbol{v}\boldsymbol{g})} \frac{\mathrm{e}^{\mathrm{i} q_{0g} L/\gamma_1} - 1}{q_{0g}} ;$$

$$\boldsymbol{v}_g = \left( \boldsymbol{v} + \frac{c\boldsymbol{g}}{\omega} \right); \qquad q_{0g} = [\omega - (\boldsymbol{v}\boldsymbol{k}_{-g})]/v; \qquad \boldsymbol{k}_{-g} = \boldsymbol{k} - \boldsymbol{g} ;$$

$$q_{\mu g s} = [\omega - (\boldsymbol{v}\boldsymbol{k}_{-g\mu s})]/v; \qquad \boldsymbol{k}_{-g\mu s} = \boldsymbol{k} - \boldsymbol{g} + \omega\epsilon_{\mu s}\boldsymbol{N}/\gamma_0 c . \quad (3.19)$$

Here the polarizations $s = 1, 2$ are assumed to be separated, and the terms $O(U/E)$ and $O(\hbar g/E)$ are neglected in the electron wavefunctions. The resulting cross-section depends essentially on the parameters $L_{0g} = 1/q_{0g}$ and

$L_{\mu gs} = 1/q_{\mu gs}$, which are the radiation coherent lengths for vacuum and crystals, respectively, and the PXR contribution to (3.19) contains the coherent superposition of both. In some works [3], the contribution proportional to $L_{0g}$ is considered as a special radiation mechanism: diffraction transition radiation. In our opinion, this separation is not sufficiently justified by the essential interference of several amplitudes in Fig. 3.3. The matrix element $M_{\rm PXR}$ is thus considered as the amplitude of the diffraction of the emitted photons in a crystal. Equation (3.19) shows that the total radiation intensity is composed of the interfering PXR and CBS amplitudes. This phenomenon was first discussed in [6] and compared with the experiment in [13]. The PXR and CBS interference is shown [10] to be critically essential in the case of nonrelativistic electrons, when the angular distributions for both radiation mechanisms are indistinguishable.

For ultrarelativistic electrons, the contributions of both intensities can be estimated in the following way. The amplitude $M_{\rm PXR}$ is most essential in the spectral $\Delta\omega/\omega$ and angular $\Delta\theta$ intervals, where the wave vector of the emitted photon satisfies the Bragg condition. In this case, the parameters in (3.15) and (3.16) are,

$$\alpha_{\rm B} \simeq \Delta\omega/\omega \simeq |\chi_{\boldsymbol{h}}|; \quad |\gamma^{h}_{\mu s}| \simeq 1\,.$$

Thus, PXR contribution within these narrow intervals has a magnitude

$$\left(\frac{\partial^2 N}{\partial\omega\partial\boldsymbol{n}}\right)_{\rm PXR} \approx \frac{\alpha}{4\pi c^4}\frac{J}{e}(\boldsymbol{e}_{gs}\boldsymbol{v}_g)^2\omega L_{0g}^2\,, \tag{3.20}$$

whereas the PXR contribution to the integral intensity of reflection is [10]

$$N_{\rm PXR} \approx \frac{\alpha}{4\pi}\frac{J}{e}\frac{|\chi_g|^2}{\sin^2\theta_{\rm B}}\frac{\omega_{\rm B}L_{\rm abs}}{c}\left[1-{\rm e}^{-L/L_{\rm abs}}\right]\,. \tag{3.21}$$

Here $L_{\rm abs}$ is the crystal absorption length for the X-rays of frequency $\omega_{\rm B} = cg/2\sin\theta_{\rm B}$, which corresponds to the Bragg frequency in the PXR reflection. The same estimations for the CBS contribution are

$$\left(\frac{\partial^2 N}{\partial\omega\partial\boldsymbol{n}}\right)_{\rm CBS} \approx \frac{\alpha}{4\pi c^4}\frac{J}{e}(\boldsymbol{e}_{gs}\boldsymbol{v}_g)^2\omega L_{0g}^2\left|\frac{U_{\boldsymbol{g}}}{\hbar cg}\right|^2; \qquad \left|\frac{U_{\boldsymbol{g}}}{\hbar g}\right| \le 10^{-2}\,;$$

$$N_{\rm CBS} \approx \frac{\alpha}{4\pi c^4}\frac{J}{e}\frac{m^2c^4}{E^2}\left|\frac{U_{\boldsymbol{g}}}{\hbar cg}\right|^2\omega L_{\rm abs}\left[1-{\rm e}^{-L/L_{\rm abs}}\right]\,. \tag{3.22}$$

Thus, the maximal spectral intensity of PXR is several orders higher than the intensity of CBS; however, CBS gives a noticeable contribution to the integral intensity of the reflection. Using (3.4) and (3.6), the ratio of CBS and PXR intensities in the reflection is

$$\xi = \frac{N_{\rm CBS}}{N_{\rm PXR}} \approx \left[\frac{Z-F(\boldsymbol{g})}{F(\boldsymbol{g})}\right]^2\left(\frac{m^2c^3}{E\hbar g}\right)^2\frac{1}{16\sin^4\theta_{\rm B}}\,, \tag{3.23}$$

i.e. CBS contributes mainly in the range

$$E \leq \left| \frac{Z - F(\boldsymbol{g})}{F(\boldsymbol{g})} \right| \left( \frac{m^2 c^3}{4\hbar g \sin^2 \theta_{\mathrm{B}}} \right) . \tag{3.24}$$

The validity of this estimation is confirmed in [13], where CBS at high energies ($E \geq 50\,\mathrm{MeV}$) is found to be negligible, except for the high-order harmonics case when the atomic scattering factor $F(\boldsymbol{g})$ is exponentially small. Figure 3.4 shows the CBS contribution to the integral intensity of the reflection, simulated by (3.19) with (dotted lines) and without (solid lines) $M_{\mathrm{CBS}}$ for experimental conditions of [3]. The intensity in Fig. 3.4 is calculated in the Bragg backward geometry as a function of the angle $\psi$ between the electron velocity $\boldsymbol{v}$ and the crystal surface normal $\boldsymbol{N}$ in the vicinity of $\psi = 0$. The effect of CBS is negligible for the electron energy $E = 85\,\mathrm{MeV}$; however, it becomes evident for $E = 30\,\mathrm{MeV}$ and essential for $E = 10\,\mathrm{MeV}$.

## 3.4 Dynamical Diffraction Effects in High-Resolution Parametric X-ray Radiation

In this section, the dynamical effects followed from (3.19) are analysed. For the narrow spectral–angular range, where these effects become apparent, the CBS matrix element can be omitted. For convenience, a special coordinate system is introduced for the PXR reflection $\boldsymbol{g}$, where the frequency $\omega$ and the angles $\theta, \varphi$ of the emitted photons are counted relatively to the vector $\boldsymbol{k}_{\mathrm{B}}$ for the exact Bragg condition $\alpha_{\mathrm{B}} = 0$. In these coordinates, the photon wave vector $\boldsymbol{k}$ with an accuracy $|\chi_{\boldsymbol{g}}| \leq 10^{-5}$ is [11]

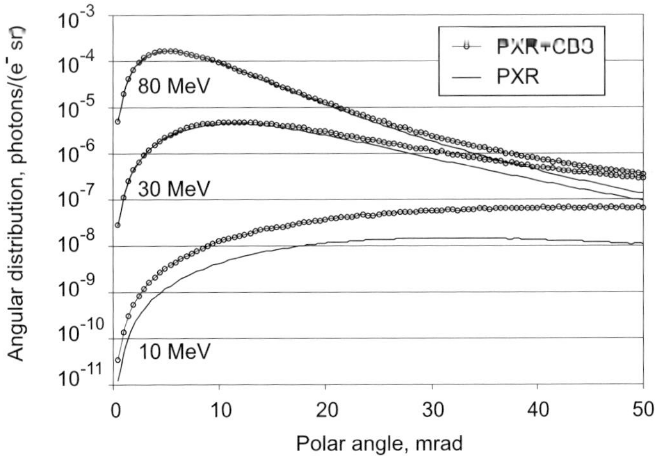


**Fig. 3.4.** Contribution of CBS to the radiation intensity dependent on the electron energy

$$\boldsymbol{k} = \boldsymbol{k}_{\mathrm{B}} + \delta\boldsymbol{k}; \qquad \delta\boldsymbol{k} = \delta\omega\boldsymbol{n}_{\mathrm{B}} + \omega_{\mathrm{B}}\theta(\boldsymbol{n}_x\cos\varphi + \boldsymbol{n}_y\sin\varphi)\;;$$
$$\delta\omega = \omega - \omega_{\mathrm{B}}; \qquad \boldsymbol{k}_{\mathrm{B}} = \omega_{\mathrm{B}}\frac{\boldsymbol{v}}{c^2} + \boldsymbol{g}; \qquad \omega_{\mathrm{B}} = \frac{hc^2}{2v|\sin\theta_{\mathrm{B}}|}\;;$$
$$\boldsymbol{n}_{\mathrm{B}} = \frac{\boldsymbol{k}_{\mathrm{B}}}{k_{\mathrm{B}}}; \qquad \theta_{\mathrm{B}} = \frac{1}{2}\widehat{\boldsymbol{v}\boldsymbol{n}_{\mathrm{B}}}\;, \tag{3.25}$$

where the unit vectors $\boldsymbol{n}_x \parallel [\boldsymbol{v}\boldsymbol{h}]$, $\boldsymbol{n}_y \parallel [\boldsymbol{n}_x\boldsymbol{n}_{\mathrm{B}}]$ lie on the plane perpendicular to $\boldsymbol{n}_{\mathrm{B}}$. The terms in (3.19) become

$$\alpha_{\mathrm{B}} = \frac{4\sin^2\theta_{\mathrm{B}}}{\cos 2\theta_{\mathrm{B}}}\frac{\delta\omega}{\omega_{\mathrm{B}}} - 2\theta\sin\varphi\tan 2\theta_{\mathrm{B}}; \qquad \gamma_0 = (\boldsymbol{v}\boldsymbol{N}); \qquad \gamma_1 = (\boldsymbol{n}_{\mathrm{B}}\boldsymbol{N})\;;$$
$$q_{0h} = \frac{1}{2}\frac{\omega_{\mathrm{B}}}{c}\left[\theta^2 + \theta^2_{\mathrm{ph}}\right]; \qquad q_{\mu hs} = \frac{1}{2}\frac{\omega_{\mathrm{B}}}{c}\left[\theta^2 + \theta^2_{\mathrm{ph}} - 2\epsilon_{\mu s}\right]\;;$$
$$\frac{(\boldsymbol{e}_{gs}\boldsymbol{v}_g)^2}{c^2} = \theta^2\nu_{\mathrm{s}}; \quad \nu_{1(2)} = \sin^2\varphi(\cos^2\varphi); \quad \theta^2_{\mathrm{ph}} = \frac{m^2}{E^2} + \theta^2_{\mathrm{sc}} + \theta^2_{\mathrm{M}}\,. \tag{3.26}$$

Similar to the above-considered PXR in thin crystals, the parameter $\theta^2_{\mathrm{ph}}$ is introduced, which takes into account the electron multiple scattering $\theta^2_{\mathrm{sc}}$ and the crystal mosaicity $\theta^2_{\mathrm{M}}$. For the crystal thickness $L < L_{\mathrm{ext}} \simeq (\omega_{\mathrm{B}}\chi_0)^{-1}$, equation (3.19) is reduced to the kinematic diffraction formula, analysed already in Chap. 2. Therefore, we consider here the domain $L > L_{\mathrm{ext}}$, where the dynamical effects are expected to be strong. This domain can be named high-resolution PXR (HRPXR) by analogy with the known high-resolution X-ray diffraction (HXRXD) field [2]. In order to emphasize the key features of the HRPXR, the case of the thick crystal $L > L_{\mathrm{abs}} \simeq (2\omega_{\mathrm{B}}\mathrm{Im}\,\chi_0)^{-1}$ is considered, where the oscillations are absent in diffraction peaks [2]. Under this assumption, the only root $\mathrm{Im}\,\epsilon_{\mu s} > 0$ in (3.19) is essential and the expression is simplified:

$$\frac{\partial^2 N_{gs}}{\partial\omega\partial\Omega} = \frac{\alpha\nu_{\mathrm{s}}\theta^2}{\omega_{\mathrm{B}}\pi^2}\left|\frac{\beta C_{\mathrm{s}}\chi_{\boldsymbol{g}}}{2\epsilon_{1(2)} - \chi_0}\right|^2\left|\frac{1}{\theta^2 + \theta^2_{\mathrm{ph}}} - \frac{1}{\theta^2 + \theta^2_{\mathrm{ph}} - 2\epsilon_{1(2)s}}\right|^2, \tag{3.27}$$

where the index 1(2) for $\omega, \theta$ is chosen depending on the sign of $\mathrm{Im}\,\epsilon_{\mu s}$. The analytical integration over the azimuth angle $\varphi$ cannot be performed in general due to the dependency $\alpha_{\mathrm{B}}(\varphi)$. There are two essentially different angular and spectral scales for characterization of the PXR reflection, as follows from (3.27). The first scale, low-resolution scale (LRS), is defined by the angular dependence of the radiation coherent length:

$$(\delta\theta)^2 \le |\chi_{\boldsymbol{g}}|; \qquad (\delta\theta)_{\mathrm{LRS}} \le \sqrt{|\chi_{\boldsymbol{g}}|} \sim 10^{-2}\text{–}10^{-3}\,. \tag{3.28}$$

The dimensionless angular and frequency distributions, obtained by integration of (3.27) over $\theta$ or $\omega$, have been kinematically considered above and investigated in numerous experiments. These distributions do not actually show any dynamical effects and their amplitudes depend on the value $L_{\mathrm{abs}}$.

The same situation occurs in the conventional low-resolution X-ray diffraction, where the reflection intensity is proportional to $L_{\mathrm{abs}}$ [2].

The high-resolution scale (HRS) in (3.27) is defined by the Bragg condition for the emitted photons:

$$|\alpha_{\mathrm{B}}| \leq |\chi_{\boldsymbol{g}}|; \qquad (\delta\theta)_{\mathrm{HRS}} \simeq \frac{\delta\omega}{\omega_{\mathrm{B}}} \leq |\chi_{\boldsymbol{g}}| \sim 10^{-5}\text{–}10^{-6}\,. \tag{3.29}$$

The experimental set-up for investigation of this fine structure of the PXR reflection is based on the double-crystal detection scheme, which is widely used in modern HRXRD, and has been recently applied for HRPXR [3].

To analyse (3.27) in the way combining the advantages of the kinematic PXR and dynamical X-ray diffraction [2] analysis, we present this distribution in the universal form, introducing new variables:

$$\eta_{\mathrm{s}} = \frac{-\beta\alpha_{\mathrm{B}} + \chi_0(\beta - 1)}{\kappa_{\mathrm{s}}}; \quad x_{\mathrm{s}} = \frac{\theta^2 + \theta_{\mathrm{ph}}^2}{\kappa_{\mathrm{s}}}; \quad \kappa_{\mathrm{s}} = 2C_{\mathrm{s}}\sqrt{|\beta|\chi_{\boldsymbol{g}}\chi_{-\boldsymbol{g}}}. \tag{3.30}$$

Then the PXR cross-section is

$$\frac{\partial^3 N_{gs}}{\partial\eta_{\mathrm{s}}\partial x_{\mathrm{s}}\partial\varphi} = \frac{\alpha\nu_{\mathrm{s}}^2}{4\pi^2\beta\sin\theta_{\mathrm{B}}} I(\eta_{\mathrm{s}}, x_{\mathrm{s}})\,, \tag{3.31}$$

with the universal function $I(\eta, x)$ for both polarizations

$$I = \frac{x - u_{\mathrm{ph}}}{x^2|\eta + \mathrm{sign}\,(\eta)\sqrt{\eta^2 - 1}|^2}\left|1 - \frac{x}{x - \eta - \mathrm{sign}\,(\eta)\sqrt{\eta^2 - 1} - u_0}\right|^2;$$
$$u_{\mathrm{ph}} = \frac{\theta_{\mathrm{ph}}^2}{\kappa_{\mathrm{s}}}; \quad u_0 = \frac{\chi_0}{\kappa_{\mathrm{s}}}; \quad x > u_{\mathrm{ph}}; \quad -\infty < \eta < \infty. \tag{3.32}$$

Figure 3.5 shows the universal function $I(\eta, x)$ with a typical HRXRD Darwin curve shape [2], caused by the factor in (3.32). This effect can be interpreted as the dynamical Bragg diffraction of the pseudophotons. Besides, a narrow and high intensity Cherenkov peak appears on the plot due to the last term in (3.32). Both processes interfere and principally are not separated into different radiation mechanisms due to their intermixed dependence on $x, u_{\mathrm{ph}}$ and $u_0$.

To illustrate the close analogy of HRXRD and HRPXR, two techniques for description of PXR are discussed below, which are widely used in modern diffractometry. The first deals with DuMond diagrams [8], which are useful for consideration of optical properties of the crystal, where the PXR is generated.

Figure 3.6a shows the dependence of the wavelength $\lambda_{\mathrm{B}}$ on the angle $\theta_{\mathrm{B}}$ between the electron velocity and crystallographic planes in a single PXR peak. Figure 3.6b for real space geometry allows us to evaluate the width of the PXR peak from the parameter $\theta_{\mathrm{ph}}$, which is the effective distribution of electron incidence angles.

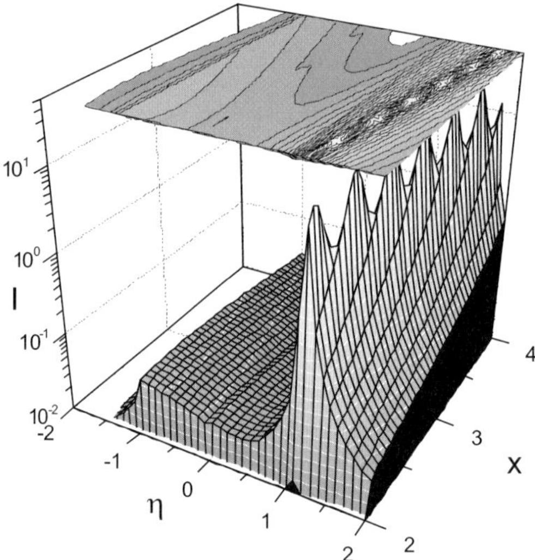


**Fig. 3.5.** Universal function for the HRPXR spectral–angular distribution ($u_{\text{ph}} = 1.5$; $u_0 = 1.0 - \text{i}\, 0.1$)

To study a fine structure of the spectral–angular distribution of PXR, the second mapping technique is convenient, which gives a two-dimensional representation of the PXR intensity as a function of two scanning variables [16], $x$ and $\eta$. The map on the top of Fig. 3.5 clearly shows the relation between the intensity and the relative width of different contributions in the PXR reflection. Estimation of the maximal the PXR spectral intensity from one electron, corresponding to the Cherenkov peak is

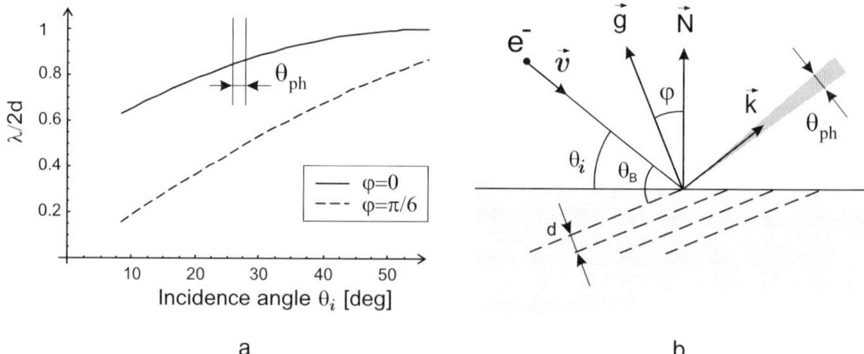


**Fig. 3.6.** Si (111) PXR reflection: (**a**) DuMond diagrams for the angle–wavelength coupling, (**b**) the corresponding real space geometry

$$\left(\frac{\partial^3 N_{gs}}{\partial\eta_s\partial x_s\partial\varphi}\right)_{\max} \simeq \frac{\alpha\nu_s^2}{4\pi^2\beta\sin\theta_B}Q^2 \; ;$$

$$Q = \frac{\mathrm{Re}\;\chi_0}{\mathrm{Im}\;(\chi_0-\kappa_s)} \sim 100 \; , \tag{3.33}$$

which is achieved in a very narrow spectral,

$$\frac{\delta\omega}{\omega} \simeq \mathrm{Im}\;\chi_0 \sim 10^{-6}\text{–}10^{-7} \; ,$$

and angular range

$$\theta^2 \simeq \kappa_s \simeq \mathrm{Re}\;\chi_0 \sim 10^{-6}\;\mathrm{mrad}^2 \; .$$

For example, the magnitude of the PXR peak in the (224) Bragg reflection in a germanium crystal with the X-ray polarizability $\chi_h = -1.44\times10^{-5} + \mathrm{i}\;7.84\times10^{-7}, \chi_0 = -2.90\times10^{-5} + \mathrm{i}\;8.65\times10^{-7}$ is:

$$I^{(s)}_{\{224\}} \approx \alpha\frac{J}{e}\frac{|\chi_{g_m}(\omega_B^{(m)})|^2}{\mathrm{Im}\;(\chi_0-\kappa_s)} \approx 5.1\times10^{18} J[\mathrm{A}]\frac{\mathrm{photons}}{\mathrm{smrad}^2(0.1\%\;\mathrm{bandwidth})}. \tag{3.34}$$

Thus, the brightness of PXR in thick crystals is higher than the one in thin films (2.48).

## References


1. A.I. Akhiezer, V.B. Berestetzkii: *Quantum Electrodynamics* (Nauka, Moscow 1969) pp 463–467
2. A. Authier: *Dynamical Theory of X-ray Diffraction* (Oxford University Press, New York 2001) pp 3–26
3. H. Backe, C. Ay, N. Clawiter, Th. Doerk, M. El-Ghazaly, K.-H. Kaiser, O. Kettig, G. Kube, F. Hagenbuck, W. Lauth, A. Rueda, A. Sharafutdinov, D. Schroff, T. Weber: *Proc. Int. Symp. on Channelling – Bent Crystals – Radiation Processes* (J.W. Goethe University, Frankfurt am Main 2003), p 41
4. V.G. Baryshevsky: *Channelling, Radiation and Reactions at Crystals under High Energy* (Belarussian State University, Minsk 1982)
5. V.G. Baryshevsky, I.D. Feranchuk: J. Phys. (Paris) **44**, 913 (1983)
6. S.V. Blazhevich, G.L. Bochek, V.B. Gavrikov, V.I. Kulibaba, N.I. Maslov, N.N. Nasonov, B.H. Pyrogov, A.G. Safronov and A.V. Torgovkin: Phys. Lett. A **195**, 210 (1994)
7. M. Born, E. Wolf: *Principles of Optics* (Pergamon Elmsford, New York 1975) p 328
8. J.W.M. DuMond: Phys. Rev. **52**, 872 (1937)
9. I.D. Feranchuk: Coherent phenomena in the processes of X-ray and gamma-radiation from relativistic charged particles in crystals. Habilitation Doctor Thesis, Belarussian State University, Minsk (1984)
10. I.D. Feranchuk, A.P. Ulyanenkov, J. Harada, J.C.H. Spence, Phys. Rev. E **62**, 4225 (2000)

11. I.D. Feranchuk, A.V. Ivashin: J. Phys. (Paris) **46**, 1981 (1985)
12. R. Jenkins, R.L. Snyder: *Introduction to X-ray Powder Diffractometry* (Wiley, New York 1996)
13. V.V. Morokhovskyi, J. Freundenberger, H. Genz, V.L. Morokhovskii, A. Richter, J.P.F. Sellschop: Phys. Rev. B **61**, 3347 (2000)
14. H. Nitta: Phys. Rev. B **45**, 7621 (1992)
15. H. Nitta: J. Phys. Soc. Japan **69**, 3462 (2000)
16. U. Pietsch, V. Holy, T. Baumbach: *High-Resolution X-ray Scattering. From Thin Films to Lateral Nanostructures* (Springer, New York 2004)
17. M.L. Ter-Mikaelian: *High Energy Electromagnetic Processes in Condensed Media* (in Russian: AN ArmSSR, Yerevan 1969) (in English: Wiley, New York 1972)

# 4

# PXR from Nonrelativistic Electrons

## 4.1 Qualitative Analysis

The observation of parametric X-ray radiation (PXR) from ultrarelativistic electrons with energy $E \geq 50\,\mathrm{MeV}$, considered in the preceding chapters, requires the use of linear particle accelerators, which usually have a low current $J \leq 10^{-7}$ A. The PXR experiments at accelerator facilities are of interest for studying the fundamental features of the phenomenon and also for several specific applications. Despite high spectral density of PXR from ultrarelativistic electrons, the parametric X-rays, being used as an X-ray source, cannot compete with modern synchrotron sources utilizing the storage rings with a current $J \simeq 1{-}10$ A [13].

Contrary to the Vavilov–Cherenkov radiation, the PXR dependence on the particle energy does not have a threshold, and therefore the radiation is still intense for electrons with $E \simeq 0.1{-}10\,\mathrm{MeV}$. For this energy, the electron beams of current $J \simeq 0.1{-}1$ A can be obtained under laboratory conditions, which opens the possibility for wide PXR applications.

The qualitative properties of PXR from nonrelativistic electrons are obviously explained using a concept of pseudophotons, introduced in Chap. 1. The electromagnetic field of a charged particle with arbitrary energy $E$ can be represented as a beam of virtual pseudophotons with almost uniform distribution in frequency, and this beam is a potential source for the real photons with white spectrum up to the energy $\hbar\omega < E$. In the framework of this model, any radiation mechanism can be considered as the converter of the pseudophotons to the real photons with some spectral interval depending on the interaction between the particle and the external field. The general formula (2.18) allows us to find a universal estimation for the spectral–angular distribution of photons produced by the electron beam in unit time due to the interaction of latter with some kind of external field or medium:

$$\frac{\partial^2 N_{\mathrm{s}}}{\partial\omega\partial\boldsymbol{n}} \simeq \frac{\alpha}{2\pi c^4}(\boldsymbol{e}_{\mathrm{s}}\boldsymbol{v})^2 R(\omega,\boldsymbol{n},E) L^2_{\mathrm{coh}}(\omega,\boldsymbol{n},E)\omega\frac{J}{e}\ . \tag{4.1}$$

Here $\alpha$ is the fine structure constant; $\boldsymbol{e}_{\mathrm{s}}$ is the polarization vector of the photon with frequency $\omega$; the unit vector $\boldsymbol{n}$ defines the direction of photon emission; $I$ is the number of electrons, with energy $E$ and velocity $\boldsymbol{v}$, passing through the interaction region in unit time.

The dimensionless coefficient $R(\omega, \boldsymbol{n}, E)$ defines the probability of the transformation of a pseudophoton into a real photon with the wave vector $\boldsymbol{k} = \omega\boldsymbol{n}/c$. The magnitude of the probability depends on the type of radiation mechanism and satisfies the inequality

$$R(\omega, \boldsymbol{n}, E) \leq 1 \,. \tag{4.2}$$

The coherent length $L_{\mathrm{coh}}$, introduced in [4,7] for the qualitative analysis of the radiation, is the universal characteristic of any radiation process. The value of the coherent length is defined by the kinematics of the interaction between the electron and radiation field and generally can be estimated to have a lesser value from the following parameters (2.48):

$$L_{\mathrm{coh}} = \min\left\{L,\ \ L_{\mathrm{abs}} = \frac{2c}{\omega\varepsilon''},\ \ L_{\mathrm{coh}} = [p_z - p_{fz} - k_z]^{-1} \equiv q_z^{-1}\right\}\,, \tag{4.3}$$

where $q_z = p_z - p_{fz} - k_z$ is the projection of the transmitted momentum in the direction of the electron velocity; $\boldsymbol{p}$ and $\boldsymbol{p}_f(E_f)$ are the wave vectors of the electron in the initial and final states, respectively, $E_f = E - \omega$, and $L$ is the sample length along the electron trajectory. Under the condition $\omega \ll E$, the parameter $L_{\mathrm{coh}}$ is

$$L_{\mathrm{coh}} \simeq \frac{c}{\omega[1 - (\boldsymbol{v}\boldsymbol{n})/c\varepsilon']}\,,$$
$$L_{\mathrm{coh}} \simeq \frac{2c}{\omega[2(1-\varepsilon') + \gamma^{(-2)}]}\,, \qquad \text{if} \quad \gamma = \frac{E}{mc^2} \gg 1\,. \tag{4.4}$$

Here $\varepsilon'$ and $\varepsilon''$ are the real and imaginary parts of the refractive index of the medium, respectively. From the physical point of view, the value $L_{\mathrm{coh}}$ defines the length of the electron path in the medium, where the emitted photons are coherent.

Expression (4.1) shows that under the assumption of constant electron energy, the ratio of intensities from various radiation mechanisms is defined by the transformation coefficients (4.2) and the coherent length. For the Cherenkov radiation, $L_{\mathrm{coh}}$ achieves the maximal magnitude under the condition $q_z = 0$ and within the spectral interval, where the inequality $\varepsilon' > 1$ is fulfilled. For PXR, $R(\omega, \boldsymbol{n}, E)$ is equal to the coefficient of the Bragg reflection from the crystallographic planes and tends to unity for the photons with the vector $\boldsymbol{k}$ located nearby the Ewald sphere [2]. Thus, the maximum value of the PXR intensity corresponds to the case when the Bragg diffraction condition for the reciprocal vector $\boldsymbol{g}$

$$(\boldsymbol{k}+\boldsymbol{g})^2 = k^2 \tag{4.5}$$

is fulfilled simultaneously with the Cherenkov condition

$$p_z - p_{fz} - k_z = 0\ . \tag{4.6}$$

The former condition (4.5) sets the value $R$ close to unity, $R \simeq 1$, whereas (4.6) forces the value $L_{\mathrm{coh}}$ to a maximum. This situation is only possible for relativistic particles with energy $E \gg m$, and the optimal value of the electron energy $E_{\mathrm{opt}}$ is about 50 MeV [3].

If (4.6) is not fulfilled exactly but the condition $\gamma \gg 1$ is still valid, the value $L_{\mathrm{coh}}$ is proportional to $\left(E/mc^2\right)^2$ and decreases as $E^2$ with the decrease in energy. The factor $(\boldsymbol{e}_{\mathrm{s}}\boldsymbol{v})^2$, being dependent on the square of the characteristic angle of radiation, increases as $E^{-2}$, according to (2.43). Therefore, the spectral intensity of radiation depends on the particle energy as follows:

$$\frac{\partial^2 N}{\partial \boldsymbol{n} \partial \omega} \sim E^2\ . \tag{4.7}$$

This fact has been experimentally confirmed for PXR in [6]. The estimation (4.7) indicates the essential feature of the spectral intensity function: the behaviour of this function is determined solely by the kinematics of the radiation process (the coherent length and polarization factor) and not by the mechanism of radiation. As a consequence, the transformation coefficient $R(\omega, \boldsymbol{n}, E)$ remains almost constant in the wide range of the electron energy.

This result predicts the losses of the radiation intensity produced by a nonrelativistic electron with energy $E \sim 0.1$ MeV, which is achievable in X-ray tubes, by the factor $10^{-5}$–$10^{-6}$ comparing to the maximal possible value. However, this decrease can be compensated by the increase of the electron current from $J \sim 10^{-8}$–$10^{-9}$ A (linear accelerators) up to $J \sim 1$ A (achievable in X-ray tubes). Another specific feature of PXR from nonrelativistic electrons follows from the angular distribution of radiation and sets a special requirement for the detection scheme. In the case of relativistic or ultrarelativistic particles passing through a crystal, the background radiation caused by effects other than PXR (for example, coherent and incoherent bremsstrahlung) is concentrated inside a narrow cone along the direction of the particle beam (Fig. 4.1a). Therefore, the quasi-monochromatic photon beams detected in the directions of Bragg angles are due to the PXR phenomenon only. As a result, even detectors with a relatively low spectral resolution are able to record the PXR peaks in the vicinity of Bragg reflections. In the case of nonrelativistic electrons, the angular distribution of radiation yielded by all emission mechanisms, including PXR and CBS, is almost isotropic (Fig. 4.1b). Therefore, peaks caused by coherent orientational effects are observed on an intense uniform background, and their shape and amplitude are mainly defined by the spectral resolution of the detector. Thus, the relativistic factor plays a principal role in the angular distribution of the emitted photons.

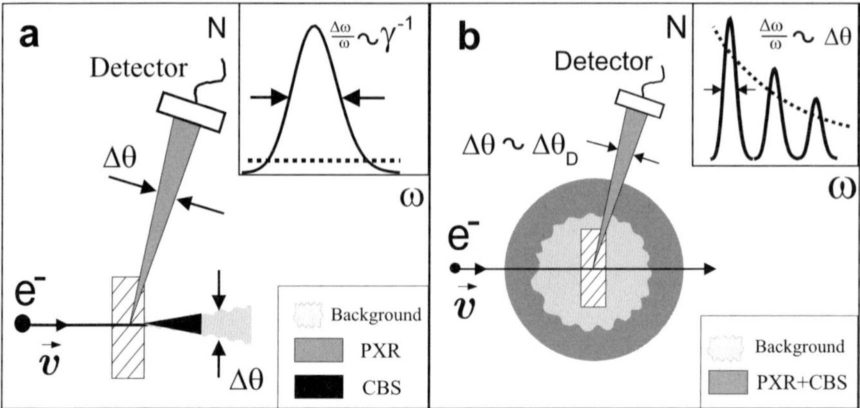


**Fig. 4.1.** The spatial distributions of the X-ray radiation intensity for relativistic (**a**) and nonrelativistic (**b**) electrons. The insets show the spectral structure of the detected peaks

Besides, in the nonrelativistic case, the electron multiple scattering in a crystal essentially influences the radiation spectral width. According to (1.26), for the electron energy $E \simeq 0.1$ MeV, the crystal length optimal for PXR output is $L \sim 0.1{-}1\,\mu\text{m}$, which is typical for the transmission electron microscopy [14]. One more physical effect, gaining strength in the nonrelativistic case, is an interference of PXR and CBS amplitudes. This effect becomes theoretically evident if both radiation modes are described by $S$-matrices of quantum electrodynamics (2.48). The radiation kinematics of relativistic particles suppresses the interference between PXR and other radiations due to the large difference between pseudophoton frequencies (1.16). Moreover, the CBS amplitude is considerably smaller than the PXR amplitude (3.23). Both these factors point to the maximal interference between PXR and CBS in the case of nonrelativistic electrons.

## 4.2 Spectral–Angular Distribution of the Radiation from Nonrelativistic Electrons in Crystals

A rigorous expression (3.19) for the cross-section of electrodynamic processes in the crystal has been derived above in the framework of quantum electrodynamics, which takes into account the coherent interaction of both electrons and photons with the media. However, it makes sense to use a general equation only if (i) the recoil effect is taken into account in radiation of hard photons, (ii) the quantization of the electron transverse motion is considered (channelling radiation). Both reasons are important in the case of relativistic electrons but are negligible in the nonrelativistic case. Therefore, the classical expression (2.16) can be used to simplify a physical interpretation of the results. Thus, the following formula for the spectral density of the photon number emitted in the direction $\boldsymbol{n}$ is used in this section:

$$\mathrm{d}^2 N^{(s)}_{\boldsymbol{n}\omega} = \frac{\alpha\omega}{4\pi^2 c^2} \left| \int_0^{t_L} \mathrm{d}t \; \boldsymbol{v}(t) \boldsymbol{E}_{\boldsymbol{k}s}(\boldsymbol{r}(t), \omega) \mathrm{e}^{-\mathrm{i}\omega t} \right|^2 \mathrm{d}\omega \mathrm{d}\boldsymbol{n} \; . \tag{4.8}$$

Here $\boldsymbol{r}(t)$ and $\boldsymbol{v}(t)$ are the coordinates and the velocity of the electron in the crystal, respectively; $\boldsymbol{E}_{\boldsymbol{k}s}(\boldsymbol{r}(t), \omega)$ is the wave field of the emitted electromagnetic wave. The intervals $\mathrm{d}\omega$ and $\mathrm{d}\boldsymbol{n}$ define the spectral and angular area, respectively, where the photons are detected, and $t_L$ represents the time over which the electron moves inside the crystal of thickness $L$. The use of thin crystalline films in PXR experiments from nonrelativistic electrons allows us to simulate the functions $\boldsymbol{r}(t)$ and $\boldsymbol{E}_{\boldsymbol{k}s}(\boldsymbol{r}, \omega)$ by the perturbation theory. This technique is important for a physical interpretation of the results because (i) the contributions from different X-ray production modes to the radiation amplitude are considered in addition, (ii) the multiple scattering of electrons in the crystal is neglected.

The electromagnetic field of the emitted radiation within the limits of the perturbation theory is (see (2.25))

$$\boldsymbol{E}_{\boldsymbol{k}s}(\boldsymbol{r}, \omega) = \boldsymbol{e}_{\mathrm{s}} \mathrm{e}^{\mathrm{i}\boldsymbol{k}\boldsymbol{r}} + \sum_{\boldsymbol{g} \neq 0} \boldsymbol{E}_{\boldsymbol{g}s} \mathrm{e}^{\mathrm{i}(\boldsymbol{k}+\boldsymbol{g})\boldsymbol{r}},$$

$$\boldsymbol{E}_{\boldsymbol{g}s} = -\frac{\chi_g}{[k_g^2 - \omega^2/c^2]} [\boldsymbol{k}_g(\boldsymbol{g}\boldsymbol{e}_{\mathrm{s}}) - \frac{\omega^2}{c^2} \boldsymbol{e}_{\mathrm{s}}], \qquad \boldsymbol{k}_g = \boldsymbol{k} + \boldsymbol{g} \; , \tag{4.9}$$

where $\boldsymbol{e}_{\mathrm{s}}$ is the polarization vector; $\chi_g$ are the Fourier components of the crystal X-ray polarizability.

Formula (4.9) can be used for radiation from low-energy electrons in a crystal of arbitrary thickness because the kinematics of PXR assumes the vector $\boldsymbol{k}$ to be far from the Ewald sphere, and dynamical effects are negligible [10]. Clearly, the interaction of electrons and electromagnetic radiation with a crystal leads not only to a change of the stationary states of electromagnetic field but also to the modification of the motion law $\boldsymbol{r}(t)$ of the electron. The emission of CBS is caused by the scattering of electrons on a coherent periodic potential, which is defined by the Coulomb interaction of the particle beam with both the electron density of the crystal and that of the nuclei. This potential follows from (3.4):

$$U(\boldsymbol{r}) = \frac{1}{\Omega} \sum_{\boldsymbol{g} \neq 0} U_{\boldsymbol{g}} \mathrm{e}^{\mathrm{i}\boldsymbol{g}\boldsymbol{r}},$$

$$U_{\boldsymbol{g}} = 4\pi e \sum_i \mathrm{e}^{\mathrm{i}\boldsymbol{g}\boldsymbol{R}_i} \frac{[Z_i - F_i(\boldsymbol{g})]}{g^2} \mathrm{e}^{-W(\boldsymbol{g})}. \tag{4.10}$$

The law of motion $\boldsymbol{r}(t)$ for an electron in potential (4.10) is found by solving the Newton equations with an accuracy up to $O(U_{\boldsymbol{g}})$:

$$\boldsymbol{r}(t) = \boldsymbol{r}_0 + \boldsymbol{v}_0 t + \boldsymbol{r}_1(t) \,,$$
$$\boldsymbol{r}_1(t) = \mathrm{i}\frac{e}{m}\sum_{\boldsymbol{g}} \frac{\boldsymbol{g}}{(\boldsymbol{g}\boldsymbol{v}_0)^2} U_{\boldsymbol{g}} \mathrm{e}^{\mathrm{i}\boldsymbol{g}\boldsymbol{v}_0 t} \,. \tag{4.11}$$

The velocity of the electron beam in vacuum is designated here as $\boldsymbol{v}_0$. The inequality $r_1 \ll v_0 t$ defines the validity region of the subsequent calculations of the radiation intensity. Substituting (4.11) into this inequality, the expression for the range of validity is straightforward:

$$\frac{4\pi\alpha Z_{\mathrm{eff}}}{mv^2 g^2 \Omega} \hbar\omega L \ll 1 \,. \tag{4.12}$$

Here $Z_{\mathrm{eff}}$ is an averaged electric charge per single atom, $\Omega$ is the volume of the crystallographic unit cell. In fact, condition (4.12) means that a restriction to the length of a particle trajectory within the crystal has the same value as the extinction length of the emitted photons, and therefore the classic approach can be applied to relatively thick crystals. The wavefunction of a particle, being a part of the matrix element for the radiation intensity, is determined by the essentially lesser extinction length of the electrons. For a crystal of thickness not exceeding the photon extinction length, the considerable phase oscillations of the electron wavefunctions are mutually cancelled in the initial and final states of the particle. Therefore, it is not necessary to take into account the multi-wave diffraction of electrons [14] for calculation of the radiation matrix elements in (2.48).

After the substitution of (4.11) into (4.8), an expression for the radiation intensity in a thin crystal is derived with an accuracy up to $O(U_{\boldsymbol{g}})$:

$$\frac{\partial^2 N^{(s)}_{\boldsymbol{n},\omega}}{\partial\omega\partial\boldsymbol{n}} = \frac{\alpha}{4\pi^2 c^2}\omega \sum_{\boldsymbol{g}\neq 0} |A_{\boldsymbol{g}s}(\omega, \boldsymbol{n})|^2 \,, \tag{4.13}$$

where the amplitudes $A_{\boldsymbol{g}s}$ are (the $z$-axis is parallel to the velocity vector $\boldsymbol{v}_0$)

$$A_{\boldsymbol{g}s} = \left\{ \boldsymbol{v}_0 \boldsymbol{E}_{\boldsymbol{g}s} - \frac{e}{m}\frac{U_{\boldsymbol{g}}}{\boldsymbol{g}\boldsymbol{v}_0}\left[ \boldsymbol{e}_s \boldsymbol{g} + (\boldsymbol{e}_s \boldsymbol{v}_0)\frac{\boldsymbol{k}\boldsymbol{g}}{g_s \boldsymbol{v}_0} \right] \right\} Q \,,$$
$$Q = \frac{\sin qL_z/v_0}{q}, \qquad q = \frac{\omega - \boldsymbol{v}_0(\boldsymbol{k} + \boldsymbol{g})}{2} \,. \tag{4.14}$$

The first term in formula (4.14) describes PXR, whereas the second term determines the coherent bremsstrahlung. The position of the intensity peaks in formula (4.14) is defined by the same kinematic factor $|Q|^2$ for both mechanisms. This factor is due to the coherent interference of radiation formed by different crystallographic planes. A similar factor (2.36) has been considered for the kinematic model of PXR from relativistic electrons in thin crystals, and in the same way, the contribution of this term to the intensity can be represented as

$$|Q|^2 = 2\pi \frac{L_{\rm abs}}{v_0} \left[1 - {\rm e}^{-L_z/L_{\rm abs}}\right] \delta(\omega - \boldsymbol{v}_0(\boldsymbol{k}+\boldsymbol{g})) \,. \tag{4.15}$$

Here $L_{\rm abs}$ is the absorption length of the crystal for X-rays of frequency determined from the zeros of the $\delta$-function argument in (4.15). Thus, the spectral and angular distributions of radiation, which result from coherent processes initiated by a nonrelativistic electron beam inside a thin crystal, are defined by the sum of resonant terms. The most important point is that both terms, i.e. parametric X-ray radiation and coherent bremsstrahlung, impose the same conditions on the frequency and the direction of the emitted photons. For each selected crystallographic reflection with the interplanar distance $d$, a set of narrow spectral lines with frequencies $\omega_n(\theta)$ is formed in the direction of the observation angle $\theta$ to the velocity vector of the electron:

$$\omega_n(\theta) = \frac{2\pi v_0 \sin\theta_{\rm B}}{d(1 - v_0/c\cos\theta)} n, \qquad n = 1, 2, \ldots \,. \tag{4.16}$$

The relative width of these lines is

$$\frac{\Delta\omega_0}{\omega} \approx \frac{v_0}{L_z \omega_n(\theta)} \,, \tag{4.17}$$

where $\theta_{\rm B}$ is the angle between the velocity $\boldsymbol{v}_0$ and the crystallographic planes scattering the X-rays. For nonrelativistic electrons, the number of emitted photons depends weakly on the variation of angle $\theta$ and is defined by the sum of the interfering amplitudes of PXR and CBS. The number of photons emitted in a chosen direction and integrated in the vicinity of the peak and in the frequency interval $\Delta\omega > \Delta\omega_0$ is expressed for $L_z \leq L_{\rm abs}$ as

$$\frac{\partial N_{\rm s}}{\partial \boldsymbol{n}} \simeq \frac{\alpha}{2\pi c^2} \omega_n \frac{L_z}{v_0} \left|A_{\rm PXR} + A_{\rm CBS}\right|^2 \,, \tag{4.18}$$

whoro

$$A_{\rm PXR} = \frac{\chi_g}{[k_g^2 - \omega^2/c^2]} \left[(\boldsymbol{v}_0 \boldsymbol{k}_g)(\boldsymbol{g}\boldsymbol{e}_{\rm s}) - \frac{\omega^2}{c^2}(\boldsymbol{v}_0 \boldsymbol{e}_{\rm s})\right] , \tag{4.19}$$

$$A_{\rm CBS} = -\frac{eU_g}{m\Omega(\boldsymbol{g}\boldsymbol{v}_0)} \left[\boldsymbol{g}\boldsymbol{e}_{\rm s} + (\boldsymbol{v}_0 \boldsymbol{e}_{\rm s}) \frac{\boldsymbol{k}\boldsymbol{g}}{\boldsymbol{v}_0 \boldsymbol{g}}\right] \,. \tag{4.20}$$

In comparison to the case of relativistic particles, a new physical result follows from formulas (4.18)–(4.20), viz. considerably different ratios of PXR and CBS amplitudes in relativistic and nonrelativistic cases. For relativistic particles, this ratio depends on the electron energy (3.23), whereas for nonrelativistic particles the ratio of the first to the second term in formula (4.18) depends only on the distribution of charge density within the elementary crystal cell. This makes it necessary to take precisely into account the instrumental function of the detector and incoherent bremsstrahlung background when performing a theoretical analysis of experimental curves .

Equations (4.19) and (4.20) explain the role of the relativistic factor in the formation of coherent peaks. The expression in the denominator of formula (4.19)

$$\delta = k_g^2 - \omega_n^2/c^2 = 2\frac{\omega_n}{c}(\boldsymbol{ng}) + g^2 \simeq g^2\left(1 - \frac{v_0 \sin\theta_{\mathrm{B}}}{c(1 - v_0\cos\theta)}\right) \tag{4.21}$$

describes the deviation of the emitted photon wave vector from the Ewald sphere. For the nonrelativistic case, when the condition $1 - v_0/c \approx 1$ is fulfilled, the value of the parameter $\delta$ is approximately equal to $\delta \approx g^2$, and the contributions of PXR and CBS to coherent peaks are of the same order. As the energy of the electron $E$ increases with $v_0 \to 1$, the parameter $\delta$ becomes negligible for some angles $\theta$, and the intensity of PXR increases proportionally to $(E/m)^2$, reaching its maximum at $E_{\mathrm{opt}} \sim m/\sqrt{|\chi_g|}$ [3].

The angular distribution of the X-ray radiation scattered from the set of crystallographic planes in the nonrelativistic case is almost isotropic, and the difference between the polarization terms in (4.19) and (4.20) can be neglected. Using (4.16) for the resonant frequency, the ratio of PXR and CBS contributions to the coherent peak is obtained:

$$\delta_g = \frac{A_{\mathrm{PXR}}}{A_{\mathrm{CBS}}} \simeq \frac{\sum_i F_i(\boldsymbol{g})\mathrm{e}^{\mathrm{i}\boldsymbol{gR}_i}}{\sum_i [Z_i - F_i(\boldsymbol{g})]\mathrm{e}^{\mathrm{i}\boldsymbol{gR}_i}} \,. \tag{4.22}$$

Equation (4.22) reflects the physical nature of peak formation in the radiation spectra. PXR contributes to peaks due to the coherent scattering of emitted photons by the atomic electrons only, whereas CBS is caused by the coherent scattering of incident charged particles both by electrons and nuclei of a crystal.

Equations (4.18)–(4.20) define the position and relative amplitudes of the lines in the ideal spectra of the coherent X-ray radiation from nonrelativistic electrons in a crystal. Figure 4.2 shows the spectra simulated on the basis of the presented theory for Si (a, b), MgO (c, d) and LiF (e, f) crystals. Panels (a) and (c) correspond to the electron velocity parallel to the $\langle 111\rangle$ crystallographic axis, (b) and (d) to that parallel to $\langle 100\rangle$ and curves (e) and (f) are calculated for electrons striking the crystal perpendicular to the (110) and (100) planes, respectively. Only the reflections contributing essentially to the peak intensities are depicted. The symbols $(hkl)$ denote a single crystallographic plane, and $\{hkl\}$ means a crystal set of plane. The ideal spectra shown in Fig. 4.2 demonstrate the positions of peaks, their absolute intensities and the contribution of the PXR intensity (black part of the bar) to the full output (full bar). The panels on the left are drawn on a logarithmic scale to emphasize the intensity recession in the high-order peak series. In the panels on the right, the contributions of radiation produced by different crystallographic planes to the full spectral lines are separated. The linear intensity scale is chosen to present a real PXR contribution in the radiation yield. The crystals and experimental conditions of [9, 11] are used for simulations (the radiation peaks from nonrelativistic electrons have been reported in

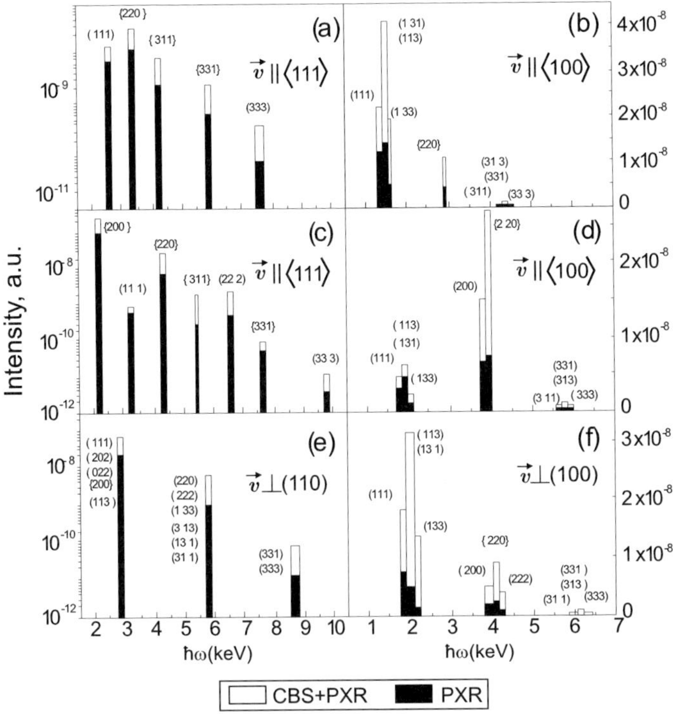


**Fig. 4.2.** The most intense series of CBS+PXR peaks for Si **(a, b)**, MgO **(c, d)** and LiF **(e, f)** crystals, for different crystal orientation. The individual reflections and sets of crystal planes contributing to the peaks are depicted near the bars

these works for the first time). The diagrams for Si and MgO are calculated for the electrons energy $E = 120$ keV and observation angle $\theta_0 = 96°$; the spectra for LiF single crystal are simulated for $E = 84$ keV and $\theta_0 = 67.5°$. The fine spectral and angular structure of peaks will be studied in the next section.

## 4.3 Simulation of Real Radiation Spectra

The principal difference between the above-derived ideal spectrum of interfering PXR and CBS from nonrelativistic particles and the PXR spectrum from ultrarelativistic particles is that the low-energy particles emit the radiation isotropically, whereas the angular distribution of the ultrarelativistic PXR represents the set of reflections with divergence determined by the relativistic factor $mc^2/E$ (Fig. 4.1a). Anothers important point is that in the relativistic case the incoherent interaction of electrons with the crystal concentrates the radiation within a narrow cone directed along the particle motion direction.

Therefore, the intensity of PXR exceeds considerably the incoherent background when the X-rays are observed in the directions other than the velocity of electrons. Under this condition, the PXR peaks can be recorded even with detectors of relatively low spectral resolution. The coherent bremsstrahlung from relativistic particles is localized in the vicinity of the electron beam direction and its spectrum shifts to the hard X-ray region.

In the experiments with nonrelativistic electrons (Fig. 4.1b), the peaks, being composed of PXR and CBS contributions, are accompanied by an intense background caused by other radiation types, e.g., the incoherent bremsstrahlung. The angular distribution of the latter is also almost isotropic for nonrelativistic particles. Therefore, the investigation of the spectral distribution of PXR+CBS peaks at fixed observation angles requires a detector with a high energy resolution, or an additional analyser crystal in front of the detector. Actually, such analysers are also necessary when the synchrotron radiation is used as an X-ray source [8].

To define the requirements for the detector, the spectral intensities of coherent X-ray radiation (CXR) and incoherent bremsstrahlung (IBS) have to be estimated. Comparing the intensities, we disregard the kinematic factors related to the weak dependence of intensity on the radiation angle, i.e., the angular distributions of both CXR and IBS are assumed to be isotropic. This allows us to determine the most essential dimensionless parameters affecting the ratio of PXR+CBS to IBS output. Using formula (28.4.2) from [1], the number of IBS quanta emitted within the spectral interval $\Delta\omega$ can be written as

$$\frac{\partial N_{\mathrm{s}}}{\partial \boldsymbol{n}} \simeq \frac{4\alpha}{3\pi} Z^2 \left(\frac{\alpha\hbar}{mc}\right)^2 \rho L_z \ln \frac{137}{Z^{1/3}} \frac{\Delta\omega}{\omega} , \tag{4.23}$$

where $\rho$ is the concentration of the scattering centres ($\rho = 1/\Omega$ for crystals with a single atom per elementary cell). The only photons produced by the IBS mechanism and emitted near one of the PXR peaks are considered here, and the amplitudes of PXR and CBS in (4.19) are assumed to be equal and do not depend on angles. Then, using the explicit expression (3.6) for polarizability, the estimation is

$$\frac{\partial N_{\mathrm{s}}}{\partial \boldsymbol{n}} \simeq \frac{\alpha}{2\pi} \omega_n L_z v_0 \left[\frac{4\pi\alpha\hbar}{mc\omega^2} \frac{S(\boldsymbol{g})}{\Omega} \mathrm{e}^{-W(\boldsymbol{g})}\right]^2 .$$

For the crystal with a single atom per unit cell, the structure amplitude $S(\boldsymbol{g})$ is

$$\frac{S(\boldsymbol{g})}{\Omega} \mathrm{e}^{-W(\boldsymbol{g})} \simeq \rho Z ,$$

and the ratio of PXR to IBS intensities within the limits of approximations used above is

$$\eta = \frac{[\partial N_{\mathrm{s}}/\partial \boldsymbol{n}]_{\mathrm{PXR}}}{[\partial N_{\mathrm{s}}/\partial \boldsymbol{n}]_{\mathrm{IBS}}} \simeq \frac{\rho c^2}{\omega_n^3} \frac{6\pi^2 v_0}{\ln 137/Z^{1/3}} \frac{\omega_n}{\Delta\omega} . \tag{4.24}$$

Thus, the ratio of intensities in the vicinity of the peak is determined by the coherency factor

$$\xi_n = \frac{c^3 \rho}{\omega_n^3} , \tag{4.25}$$

which results from the interference of the radiation emitted by the electrons on the periodic atomic structure of a crystal. For example, for the experiment [9] with a LiF crystal and electrons of energy 63 keV penetrating the crystal parallel to the $\langle 100 \rangle$ crystallographic axis and the observation angle $\theta = 67.5°$, the coherency factor $\xi \simeq 0.39$ for spectral series related to the (200) reflection, and $\xi \simeq 0.79$ for the (111) reflection. To distinguish spectral peaks from the uniform background, the detector has to have a resolution satisfying the condition $\xi_1 \gg 1$:

$$\xi_1 = \xi \frac{\omega}{\Delta\omega} . \tag{4.26}$$

The parameter $\xi_1$, describing the signal/background ratio in the spectrum, reaches its maximum value when the resolution of the detector is of the same order as the linewidth of PXR, as formula (4.17) states. In real experiments, this resolution value can be achieved by using an analyser crystal. Assuming the same conditions as in the experiment [9], the maximum value of the parameter $\eta$ for the reflection the (200) is approximately equal to

$$\eta_{\mathrm{max}} \simeq \frac{c^2 \rho}{\omega_n^3} \frac{6\pi}{\ln 137/Z^{1/3}} \omega_n L_z \simeq 1.3 \times 10^3 , \tag{4.27}$$

for the spectral resolution of the detector 0.1%. The resolution of detectors in the experiments [9,11] was, however, only about 10%. The better the spectral resolution of the detector, the larger the coherency factor, in accordance with (4.24), and the higher the relative magnitude of PXR peaks in comparison to the background of incoherent bremsstrahlung. This result has been recently confirmed in experiments [12], where the low-energy part of the radiation spectrum was recorded using a crystal spectrometer with a 40 eV resolution and a new fine structure of peaks has bean revealed.

The absolute number of photons emitted by one electron and contributing to the PXR+CBS peak is essentially smaller than the PXR intensity from ultrarelativistic electrons. However, due to the considerably higher current density, which can be achieved for nonrelativistic particles, the integral number of quanta is relatively large in this case. Using again the parameters of the experiment [9], i.e. the electron current and energy 1 μA and 63 keV, respectively, 100nm thick LiF crystal, velocity of electrons is parallel to the $\langle 100 \rangle$ axis, the estimation for the absolute number of detected photons in the vicinity of the peak from the (200) reflection is given by (4.18)

$$N_0 \simeq 5.1 \times 10^2 \text{ photons s}^{-1} \qquad \omega_0 \simeq 3.89 \text{ keV} . \tag{4.28}$$

Here the values for the Bragg and observation angles are taken as $\theta_{\mathrm{B}} = 0$ and $\theta = 67.5°$, respectively, and the detector registers the photons in the solid

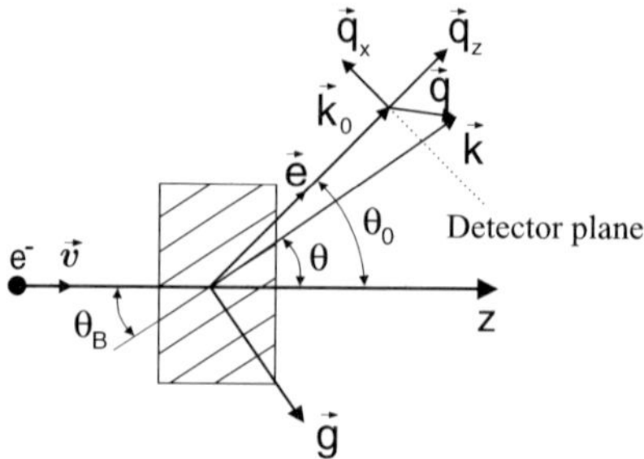


**Fig. 4.3.** Geometrical sketch of vectors and angles describing the PXR and CBS processes

angle $\Delta\Omega \sim 10^{-3}$ steradian. This estimation is in good agreement with the radiation intensity observed in [9].

Now we calculate the real PXR+CBS spectra, taking into account the convolution of the ideal spectra with both angular and spectral resolution functions of the detector and the background of incoherent radiation. To perform this convolution, the following substitution is used (Fig. 4.3):

$$\boldsymbol{k} = \boldsymbol{k}_0 + \boldsymbol{q}; \qquad \boldsymbol{k}_0 = \omega_g \boldsymbol{e} \, . \tag{4.29}$$

Here the vector $\boldsymbol{e}$ defines the direction of the detector centre and the parameter

$$\omega_g = \frac{v \sin \theta_{\mathrm{B}}}{(1 - v \cos \theta_0)} g \tag{4.30}$$

determines the frequency of a peak in the ideal PXR+CBS spectrum, corresponding to the reciprocal lattice vector $\boldsymbol{g}$. The instrumental functions describing the angular and spectral resolutions of the detector are

$$\begin{aligned} f_1(\theta) &= \frac{1}{\pi} \exp\left[-\frac{(\theta - \theta_0)^2}{\Delta\theta^2}\right] , \\ f_2(\omega) &= \frac{1}{\sqrt{\pi}} \exp\left[-\frac{(\omega - \omega_{\mathrm{r}})^2}{\Delta\omega^2}\right] , \end{aligned} \tag{4.31}$$

where $\Delta\theta$ is the angle aperture of the detector with a pin-hole slit at the front, $\Delta\omega$ is the detector spectral resolution and the frequencies $\omega_{\mathrm{r}} = r\varepsilon$ ($r = 0, 1, 2, \ldots$) correspond to the $r$th detector channel with width $\varepsilon$. Because in real experiments $\varepsilon < \Delta\omega \ll \omega_{\mathrm{r}}$ and $\Delta\theta \ll \theta_0$, the phase volume of the detected photons can be expressed via new variables as

$$\begin{aligned} \omega &\simeq \omega_g + q_z + \frac{q^2}{2\omega_0} , \\ \mathrm{d}\boldsymbol{k} = \mathrm{d}\boldsymbol{q} &\simeq \omega_{\mathrm{r}}^2 \eta \, \mathrm{d}\eta \, \mathrm{d}\varphi \, \mathrm{d}\nu , \\ \nu = \omega - \omega_{\mathrm{r}}, &\qquad \eta = \theta - \theta_0 , \end{aligned} \tag{4.32}$$

where the $z$-axis is parallel to the vector $\boldsymbol{k}_0$ and $\varphi$ is the angle between the vector $q_\perp$ and the perpendicular to the plane defined by the vectors $\boldsymbol{v}_0$ and $\boldsymbol{g}$. Then the number of counts $N_\mathrm{r}$ in $r$th channel of the detector normalized for one electron, for the conditions of the experiment [9, 11], is

$$N_\mathrm{r} = B\Delta\theta^2 \frac{\Delta\omega}{\omega_\mathrm{r}} + \frac{1}{\pi^{3/2}} \int_0^\infty \eta \mathrm{d}\eta \int_0^{2\pi} \mathrm{d}\varphi \int_{-\infty}^{\infty} \mathrm{d}\nu \left\{ \mathrm{e}^{-\frac{\eta^2}{\Delta\theta^2}} \mathrm{e}^{-\frac{(\nu-\omega_\mathrm{r})^2}{\Delta\omega^2}} \right.$$
$$\left. + \sum_{\boldsymbol{g}} A_g \delta[(\nu + \omega_\mathrm{r} - \omega_g)(1 - v_0/c \cos\theta_0) - \omega_\mathrm{r} v_0 \eta \cos\varphi] \right\}. \quad (4.33)$$

Here $A_g$ and $B$ are the amplitudes of the coherent and incoherent components of the radiation, respectively. According to formulas (4.18)–(4.23), they are

$$A_g = \frac{\alpha}{2\pi c^2} v_0 \omega_\mathrm{r} L_z |\chi_g(\omega_\mathrm{r})|^2 \left|1 + \delta_g^{-1}\right|^2, \quad (4.34)$$

$$B = \frac{4\alpha}{3\pi} Z^2 \left(\frac{e^2\hbar}{mc}\right)^2 \frac{L_z}{\Omega} \ln \frac{137}{Z^{1/3}}. \quad (4.35)$$

Performing the integration over $\nu$ in (4.33), we arrive at

$$N_\mathrm{r} = B\Delta\theta^2 \frac{\Delta\omega}{\omega_\mathrm{r}} + \frac{1}{\pi^{3/2}} \int_0^\infty \eta \mathrm{d}\eta \int_0^{2\pi} \mathrm{d}\varphi \mathrm{e}^{-\frac{\eta^2}{\Delta\theta^2}} \sum_{\boldsymbol{g}} A_g$$
$$\times \exp\left[ -\frac{(\omega_g - \omega_\mathrm{r} + u_\mathrm{r}\eta\cos\varphi)^2}{\Delta\omega^2} \right], \quad (4.36)$$

where

$$u_\mathrm{r} = \frac{v_0 \omega_\mathrm{r}}{1 - v_0/c \cos\theta_0}.$$

In general, the integrals in formula (4.36) have to be calculated numerically. However, within the limits of the approximations used in this chapter, a simple analytical formula can be derived for the radiation spectrum by using the Feynman–Jensen inequality

$$\langle \mathrm{e}^A \rangle \geq \mathrm{e}^{\langle A \rangle},$$

which is known from applications of statistical physics [5] and valid for averaging over normalized functions of a statistical distribution. As a result, the following approximate expression can be obtained:

$$N_\mathrm{r} \simeq \frac{\alpha}{2\pi} L_z \Delta\theta^2 |\chi_0(\omega_\mathrm{r})|^2 \, \omega_\mathrm{r} \left\{ \frac{1}{6\pi^2} \omega_\mathrm{r}^3 \Omega \left( \ln \frac{137}{Z^{1/3}} \right) \frac{\Delta\omega}{\omega_\mathrm{r}} \right.$$
$$\left. + v_0 \sum_{\boldsymbol{g}} \left| \frac{\chi_g(\omega_\mathrm{r})}{\chi_0(\omega_\mathrm{r})} \right|^2 \left|1 + \delta_g^{-1}\right|^2 \exp\left[ -\frac{(\omega_g - \omega_\mathrm{r})^2}{\Delta\omega^2} - \frac{\Delta\theta^2 u_\mathrm{r}^2}{4\Delta\omega^2} \right] \right\}, \quad (4.37)$$

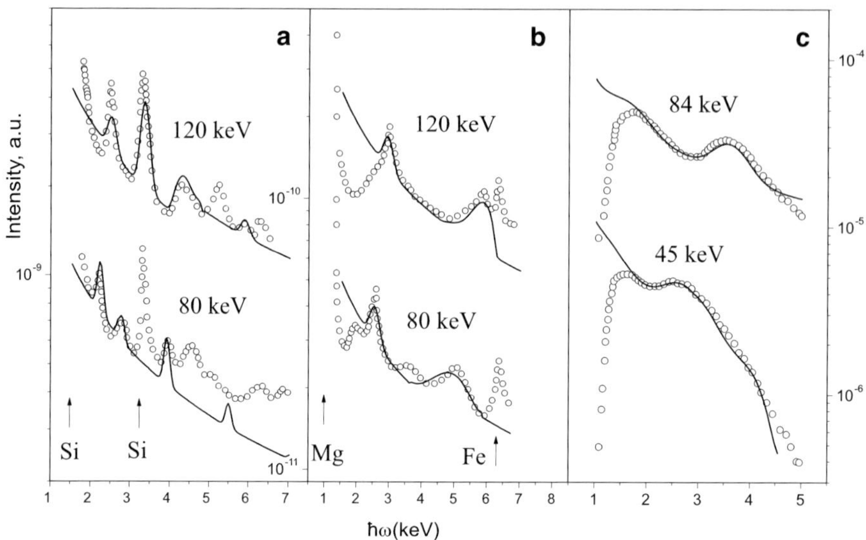


**Fig. 4.4.** Theoretical simulations (*solid lines*) and experimental data (*open dots*) reproduced from [11] for (**a**) Si and (**b**) MgO and from [9] for (**c**) LiF. The arrows show the emission lines of Si, Mg and Fe. Each panel contains two curves at different energy of electrons

which permits us to the study the dependence of the real PXR+CBS spectra on the principal parameters of the experiment. Figure 4.4a–c demonstrate the experimental data (open dots) reproduced from [9, 11] and theoretical spectra (solid lines) simulated by (4.37) for crystals Si, MgO ($\boldsymbol{v}_0 \parallel \langle 111\rangle$) and LiF ($\boldsymbol{v}_0 \perp (110)$), respectively. The parameters of the crystals and detectors are shown in Table 4.1. For each crystal, we depict two pairs of curves corresponding to different electron energies, demonstrating the dependence of the peak positions on the velocity of the electrons. The characteristic emission lines of Si, Mg and Fe are indicated by vertical arrows. The probable reasons of discrepancy between theory and experiment in the fits presented above are (i) the uncertainty of the angular resolution of detectors in experiments [9, 11], (ii) the approximate modelling of the instrumental function in (4.31) and (iii) the neglect of the background radiation from back- and multiple-scattered electrons. The disagreement of simulations and measurements in the low-frequency region is caused by the low-energy threshold of detector sensitivity, which has not been taken into consideration in calculations. The experimental results reported in [9, 11] confirm the contribution of both radiation types, coherent bremsstrahlung and parametric X-ray radiation, to the coherent yield from nonrelativistic electrons. The results of this section also indicate the possibility of qualitative and quantitative description of X-ray radiation from nonrelativistic electrons in thin crystals on

**Table 4.1.** Crystal data and parameters of the experiments [9, 11]

| Crystal | $a_0$ (Å) | $\Delta\omega$ (keV) | $\theta_0$ (deg) | $\Delta\theta^2$ (sterad) | $L_z$ (Å) | $\lvert\chi_0\rvert_{\omega=5\ \mathrm{keV}}$ |
|---|---|---|---|---|---|---|
| Si [11] | 5.4309 | 0.1 | 96 | 0.05 | 1000 | $4.03 \times 10^{-5}$ |
| MgO [11] | 4.21 | 0.1 | 96 | 0.05 | 1000 | $6.03 \times 10^{-5}$ |
| LiF [9] | 4.0276 | 0.2 | 67.5 | 0.002 | 1000 | $4.13 \times 10^{-5}$ |

the basis of the perturbation theory. The use of the classical electrodynamics approach allows us to avoid cumbersome calculations based on the multi-wave theory of electron diffraction, as treated in [11].

The experiments with a high spectral resolution are necessary for detailed studies on coherent radiation [12], which can realize the exact separation of PXR+CBS peaks from other types of radiation. For example, two- or three-crystal arrangements utilizing additional analyser crystals might be used.

## References


1. A.I. Akhiezer, V.B. Berestetzkii: *Quantum Electrodynamics* (Nauka, Moscow 1969) pp 463–467
2. A. Authier: *Dynamical Theory of X-ray Diffraction* (Oxford University Press, New York 2001) pp 3–26
3. V.G. Baryshevsky, I.D. Feranchuk: Nucl. Instrum. Methods **228**, 490 (1985)
4. B.M. Bolotovskii, G.V. Voskresenskii: Usp. Fiz. Nauk **88**, 209 (1966)
5. R.P. Feynman: *Statistical Mechanics* (W.A. Benjamin, MA 1972)
6. J. Freundenberger, V.B. Gavrikov, M. Galemann, H. Genz, L. Groening, V.L. Morokhovskii, V.V. Morokhovskii, U. Nething, A. Richter, J.P.F. Sellschop, N.F. Shul'ga: Phys. Rev. Lett. **74**, 2487 (1995)
7. V.M. Galitsky, I.I. Gurevich: Nuovo Chimento **32**, 396 (1964)
8. M. Hart: Nucl. Instrum. Methods **A 297**, 306 (1990)
9. Y.S. Korobochko, V.F. Kosmach, V.I. Mineev: Sov. Phys. JETP **21**, 834 (1965)
10. H. Nitta: Phys. Lett. A **158**, 270 (1991)
11. G.M. Reese, J.C.H. Spence, N. Yamamoto: Phil. Mag. A **49**, 697 (1984)
12. J.C.H. Spence, M. Lund: Phys. Rev. B **44**, 7054 (1991)
13. H. Wiedemann: *Particle Accelarator Physics* vol 1 & 2 (Springer, New York 1993)
14. D.B. Williams, C.B. Carter: *Transmission Electron Microscopy, vol* 1–4 (Plenum, New York 1996)

# Part II

# Experiments and Applications

# 5

# Interpretation of Experimental Results

## 5.1 Experimental Observation of PXR

Experimental studies of PXR have been performed in a wide range of electron energies. As already mentioned in Sect. 1.4, PXR was first experimentally realized in 1985. Baryshevsky and coauthors [5, 44] used 900 MeV electrons from the Sirius (Tomsk, former USSR) synchrotron and a diamond crystal. Later, numerous experiments with diamond and silicon crystals, investigating an angular [4, 6, 30, 43] and electron beam dependence [6] in the electron energy range 200–900 MeV, were reported by the same group. The facilities where PXR was measured after the Tomsk were synchrotron ARUS (Yerevan, former USSR) [26] and LINAC LUE-40 (Kharkov, former USSR) [1]. Since these experiments, linear accelerators have been extensively used for production of parametric X-rays, promising a less expensive and more affordable way to produce parametric X-rays. In the past decade, a number of LINAC facilities contributed to PXR research: Naval Postgraduate School (Monterey, USA) [49,50], S-DALINAC (Darmstadt, Germany) [53], SAL (Saskatchewan, Canada) [51], Tohoku University (Sendai, Japan) [75], Kyoto University (Kyoto, Japan) [55], Hokkaido University (Sapporo, Japan) [16], Rensselaer Polytechnic Institute (Troy, USA) [74]. The synchrotron facilities MAMI (Mainz, Germany) [40] and INS (Tokyo, Japan) [24] were utilized for PXR from ultra-relativistic electrons. The use of synchrotrons and linear accelerators allows us to cover a broad range of electron energies, from 3.5 MeV [53] to 4500 MeV [26]. Parametric X-rays were also measured from 70 GeV protons [14] at IHEP (Serpukhov, Russia) proton accelerator in 1991 and recently [3] at JINR (Dubna, Russia) from 5 GeV protons.

Both Laue (Fig. 5.1a) and Bragg (Fig. 5.1b) geometries are applied for PXR measurements; the latter is also used in backward scattering experiments (Fig. 5.1c) [27,54]. An extremely asymmetric diffraction (EAD) geometry (the Bragg-to-Laue transition case) has been also applied in several experimental set-ups [13]. In the experiment [75], the electron beam has been directed parallel to the sample surface yet not penetrating inside a crystal. The summary

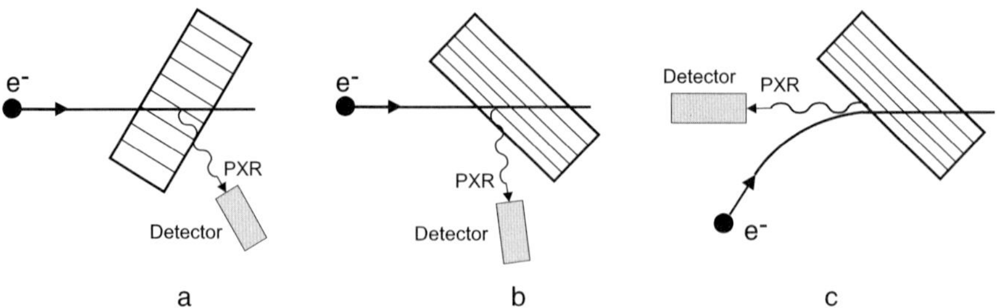


**Fig. 5.1.** Typical experimental geometries for PXR production: (**a**) Laue, (**b**) Bragg, (**c**) Bragg backwards

of basic experimental parameters of some PXR experiments is presented in Table 5.1.

For a target, the diversity of crystals has been explored. Silicon single crystal is one of the most frequently used crystals [1, 8, 13, 14, 16, 24, 27, 37, 39,40,45,51,54,55,67,70,73] with different Bragg reflections excited for PXR production: (111), (220), (333), (311), (224), etc. The other materials used for a target are diamond [5, 26, 30, 43, 44, 52, 53], LiF [74], quartz and teflon [75], GaAs [11, 12], Ge [7, 74], W [2], Cu and pyrolytic graphite [10, 50, 74]. The physical properties of the crystal, e.g. lattice period, absorption and thermal characteristics, purity, etc., influence the spectral position of the PXR reflection as well as the intensity of the peaks. For example, the PXR production competes with absorption of radiation along the escape path of photons in the crystal. For materials composed of heavy elements, the absorption losses of PXR increase more rapidly than the PXR yield, which makes light element targets more attractive. There are also numerous experiments with multilayered and grating targets [20–22, 59–61] where the principle of constructive interference of PXR and transition radiation is used to gain a radiation output.

As mentioned in Sect. 2.2, two types of detectors are used to record angular and energy dispersive dependence, respectively. For spatial resolution of PXR peaks, the proportional counter (PC) [8, 9, 24, 26, 45] or Si-PIN [16, 39, 54, 55, 70, 74] or CCD [27, 40, 67] detectors are used. In some experiments [27], the additional analysing crystal along with the CCD is used to combine an energy dispersive and angular schemes. For energy dispersive measurements of spectral PXR distributions, the NaI scintillation spectrometer [5, 8, 9, 24, 30, 44, 45] or the silicon drift Si(Li) detectors are mostly utilized [27, 52, 53]. In the work [67], a special CCD camera was used for simultaneous determination of the photon energy and position.

After the primary experiments, which discovered and confirmed the PXR phenomenon [4–6, 30, 43, 44], most of the follow-on experiments were dedicated to investigation of special features of parametric X-rays. The natural linewidth and line shape of PXR peaks were studied in [40, 52] using an absorbing foil

**Table 5.1.** Principal parameters of some published PXR experiments

| Experiment | Geometry | Crystal, $t$ | $E_{e^-/p^+}$ (MeV) | Detector |
|---|---|---|---|---|
| SIRIUS, Tomsk [5, 30, 43, 44] | Laue/Bragg | C, 0.35; 0.8 mm | 900 | NaI(Tl), PC |
| SIRIUS, Tomsk [8, 9] | Laue | Si, 0.35 mm | 900 | NaI(Tl) |
| ARUS, Yerevan [26] | Laue | C, 0.2; 1 mm | 4,500 | NaI |
| LUE-40, Kharkov [1] | Laue | Si, 30 µm | 900 | Si(Li) |
| SIRIUS, Tomsk [11, 12] | Laue | GaAs, 400 µm | 500, 900 | PC |
| INS, Tokyo [24] | Laue | Si, 0.5 mm | 200-1,100 | NaI(Tl) |
| NPS, Monterey [49, 50] | Laue/Bragg | Si, 20–320 µm; graphite, 1.39 mm | 90 | Si(Li) |
| LUE-40, Kharkov [37, 73] | Laue | Si, 20; 30 µm | 15,25 | Si(Li) |
| IHEP, Serpukhov [14] | Laue | Si, 18 mm | 70,000 ($p^+$) | $YAlO_3:Ce^{3+}$ |
| SIRIUS, Tomsk [13] | Laue/Bragg/EAD | Si, 0.15; 0.4 mm | 300–900 | PC |
| INS, Tokyo [45] | Laue | Si, 0.2–5 mm | 900 | NaI |
| S-DALINAC [52, 53] | Laue | C, 55 µm | 3.5–9.1 | Si(Li) |
| SAL, Saskatchewan [51] | Laue | Si, 20 µm | 230 | CCD |
| MAMI, Mainz [40] | Laue | Si, 124 µm | 855 | CCD |
| MAMI, Mainz [39] | Laue | Si, 100–600 µm | 855 | Si-PIN |
| S-DALINAC [67] | Laue | Si, 13 µm | 80.5 | CCD |
| ICR, Kyoto [55] | Bragg | Si, 250 µm; 3 mm | 100 | Si-PIN |
| Tohoku LINAC, Sendai [75] | surface | quartz, teflon | 150 | Si bolometer |
| S-DALINAC [54] | Bragg backwards | Si, 12.5 µm; C, 50 µm | 30.2–87 | Si-PIN |
| SIRIUS, Tomsk [61] | Bragg | $[W/B_4C]_{300}/Si$, 0.36 µm | 500 | CdTe |
| SIRIUS, Tomsk [59] | Laue | GaAs strips, 300×14 µm | 500 | CdTe |
| Hokkaido LINAC, Sapporo [16] | Laue | Si, 200–625 µm | 45 | Si-PIN |
| MAMI, Mainz [27] | Bragg backwards | Si, 525 µm | 855 | Si(Li), CCD |
| PRI, Troy [74] | Laue, Bragg | LiF, 1.5 mm | 56 | Si-PIN |
| JINR, Dubna [3] | Bragg | Si, 300 µm; graphite, 2 mm | 5,000 ($p^+$) | Si(Li) |

(Cu, Ti, Ni) technique. The PXR yield dependence on the crystal thickness was the aim of experiments reported in [39, 45, 75]. The polarization of parametric X-rays was measured in [8, 9, 67, 70], the PXR production dependence on the energy of electrons was measured in [16, 24, 53], the temperature influence on the PXR intensity was measured in [18], the effect of electron beam parameters was observed in [16, 75] and the crystal materials were optimized in [74]. The experiments [37, 64] deal with the interference of PXR and coherent bremsstrahlung. The influence of the multiple scattering on the PXR yield and radiation threshold behaviour have been investigated in [13, 24, 45]. PXR in the forward direction of electron velocity is considered in [17, 28, 58].

Up to now, the majority of PXR experiments have been carried out to study fundamental physical properties of phenomenon and to investigate principal dependencies. However, there are several works on optimization of the PXR yield for real applications. The use of PXR for mammography imaging is discussed in [68]; the application of parametric X-rays for metrology of semiconductor quantum wires and wafers is proposed in [16]; the optimization of experimental conditions to create a tunable X-ray source is performed in [23, 74]. In [31–34], the authors propose to use the forward directed and PXR-related radiations as a generation mechanism for free-electron lasers.

All above-cited experiments concern the PXR from relativistic and ultra-relativistic electrons. There are also a few experiments on the interference of PXR and CBS from nonrelativistic electrons. Korobochko et al. [62] measured the X-ray radiation from electrons (45 and 84 keV) passing through a LiF crystal. Later, similar experiments [71] were conducted with Si and MgO crystals and 40–120 keV electrons.

## 5.2 Spectral Distribution of PXR Peaks from Relativistic Electrons

The quantitative interpretation of PXR measurements depends essentially on the experimental geometry and instrumental function of a detector. To illustrate this fact, the results of pioneering experiments [5, 78] shown in Fig. 1.7 are considered below. Theoretical analysis [35,36] of these data shows that the condition of the EAD geometry is fulfilled in this case when either incident or diffracted waves travel along the crystal surface [19]. In both experiments, the electron beam with energy 900 MeV impinged on the edge of a diamond crystal, and the energy spectrum of X-ray radiation was measured at 90° to the electron velocity (Fig. 5.2a). In [5], the electron velocity was parallel to the $\langle 110\rangle$ crystallographic axis; thus the detector recorded the radiation from pseudophotons diffracted from $\{100\}$ planes of the crystal (Fig. 5.2b). The experiment [78] used the opposite geometry, i.e. $\boldsymbol{v}||\langle 100\rangle$, and PXR from $\{110\}$ was measured. In both cases, $2\theta_{\mathrm{B}} = 90°$, which corresponds to EAD.

In the case of EAD, the expression for the field in a crystal $\boldsymbol{A}^{(+)s}_{\boldsymbol{k}\omega}(\boldsymbol{r})$, required for calculation of the PXR intensity, does not follow from (3.14)–(3.16). The reason is the solutions of the dispersion equation $\epsilon_{\mu s}$, which have to be corrected along with the boundary conditions by considering the specularly

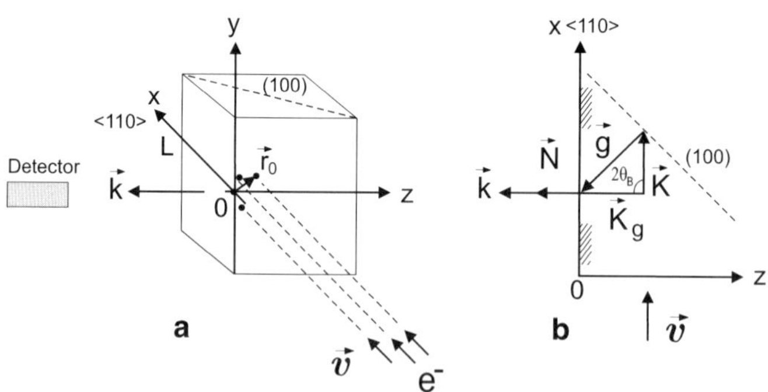


**Fig. 5.2.** (**a**) The relative position of the electron beam, crystal and detector in [5,78]; (**b**) the sketch of the diffraction process

reflected waves from the crystal surfaces [19]. To construct the solution for homogeneous Maxwell's equations in the EAD case, the wave vector $\boldsymbol{K}$ in a medium is defined in such a way that its tangential component is continuous at the boundary crystal vacuum (Fig. 5.2):

$$\boldsymbol{K} = \boldsymbol{k} - k\epsilon \boldsymbol{N} \ , \tag{5.1}$$

where $\boldsymbol{k}$, $|\boldsymbol{k}| = k = \omega/c$, is the wave vector in vacuum and the other parameters are the same as in (3.14)–(3.16). Because the vector $\boldsymbol{k}$ corresponds to the wave travelling from the detector to the radiating particle, the diffracted wave has a wave vector $\boldsymbol{k}_g = \boldsymbol{k} + \boldsymbol{g}$ along the crystal surface, and $\gamma_g = (\boldsymbol{k}_g \boldsymbol{N})/k \ll 1$. Thus, for calculation of the effective polarization $\epsilon$ in a medium, the cubic equation instead of the quadratic one has to be used:

$$-2\gamma_0\epsilon^3 + 4\gamma_0\gamma_g\epsilon^2 + 2\gamma_0(\alpha_{\mathrm{B}} + \chi_0)\epsilon + \chi_0^2 + \alpha_{\mathrm{B}}\chi_0 - C_{\mathrm{s}}^2\chi_g\chi_{-g} = 0 \ . \tag{5.2}$$

The roots of this equation have been investigated in detail in [42,57]. For the solution of the boundary condition EAD, only two of them $\epsilon_{\mu s}$ ($\mu = 1,2$), corresponding to damping inside a crystal wave, have to be used. As a result, the expression for the vector potential of a radiation field is as follows (compare with (3.14)):

$$\begin{aligned} \boldsymbol{A}_{\boldsymbol{k}\omega}^{(+)s}(\boldsymbol{r}) = \sqrt{4\pi}\Big\{ & (\boldsymbol{e}_{\mathrm{s}}\mathrm{e}^{\mathrm{i}\boldsymbol{k}\boldsymbol{r}} + A_{sp}^s \boldsymbol{e}_{gs}\mathrm{e}^{\mathrm{i}\boldsymbol{k}'\boldsymbol{r}})H(-z) \\ & + \mathrm{e}^{\mathrm{i}\boldsymbol{k}\boldsymbol{r}} \sum_{\mu=1,2} \mathrm{e}^{-\mathrm{i}k\epsilon_{\mu s}z}(\boldsymbol{e}_{\mathrm{s}}A_{\mu s} + \boldsymbol{e}_{gs}A_{g\mu s}\mathrm{e}^{\mathrm{i}\boldsymbol{g}\boldsymbol{r}})H(z)\Big\} \ , \end{aligned} \tag{5.3}$$

where $A_{sp}^s$ and $\boldsymbol{k}' = (\boldsymbol{k}_\perp + \boldsymbol{g}_\perp, \sqrt{k^2 - (\boldsymbol{k}_\perp + \boldsymbol{g}_\perp)^2})$ are the amplitude and the wave vector, respectively, for the specularly reflected wave. The following expressions for the field amplitudes can be calculated [35,36]:

$$\begin{aligned} & A_{1s} = \frac{(2\epsilon_{1s}\gamma_0 + \chi_0)(\gamma_0' - \gamma_g + \epsilon_{2s})}{(\epsilon_{2s} - \epsilon_{1s})[2\gamma_0(\gamma_0' - \gamma_g) + 2\gamma_0(\epsilon_{2s} + \epsilon_{1s}) + \chi_0]} , \qquad \gamma_0' = \frac{(\boldsymbol{k}'\boldsymbol{N})}{k} , \\ & A_{2s} = 1 - A_{1s}, \quad A_{g\mu s} = -\frac{2\epsilon_{\mu s}\gamma_0 + \chi_0}{\chi_{-g}C_{\mathrm{s}}} A_{\mu s}, \quad A_{sp}^s = A_{g1s} + A_{g2s} \ . \end{aligned} \tag{5.4}$$

Replacing $\boldsymbol{k} \to -\boldsymbol{k}$ in the wave field derived above, the matrix element (3.18) is directly calculated. Then the spectral–angular distribution of photons from a single electron in the PXR reflection $\boldsymbol{g}$ and a crystal of thickness $L$ can be simulated:

$$\frac{\partial^2 N_{gs}}{\partial\omega\partial\boldsymbol{n}} = \frac{\alpha\omega}{4\pi^2c^4}(\boldsymbol{e}_{gs}\boldsymbol{v})^2 \left| \sum_{\mu=1,2} A_{g\mu s}L_g\left(1 - \mathrm{e}^{-\mathrm{i}L/L_g}\right) \right|^2 \mathrm{e}^{-2z_0k\epsilon_{\mu s}''} \ . \tag{5.5}$$

Here $L_g = c[\omega - \boldsymbol{v}(\boldsymbol{k} + \boldsymbol{g})]^{-1}$ is a coherent radiation length, $z_0$ is the point of electron penetration inside a crystal, $\epsilon_{\mu s}''$ is an imaginary part of polarizability and $\boldsymbol{n}$ is a unit vector to the detector. For relativistic particles, the main contribution to the radiation amplitude is given by the wave field travelling along the electron trajectory, i.e., the diffracted wave (Fig. 5.2).

Two facts should be emphasized for PXR in the EAD mode. According to (5.5), the intensity decreases within the small distance from the entry point $z_0$. This is contrary to the conventional Bragg geometry (3.19), where the decrease takes place along the entire crystal length. The second point is that the coherence length is independent of the polarizability $\epsilon_{\mu s}$ due to the condition $(\boldsymbol{vN}) \approx 0$. Thus, for the EAD geometry, the 'density effect', [76], influencing the radiation processes in a homogeneous medium, disappears. For ultrarelativistic electrons, the following geometrical conventions are useful:

$$\boldsymbol{n} = -\cos\theta\ \boldsymbol{N} + \theta_x \boldsymbol{e}_x + \theta_y \boldsymbol{e}_y, \qquad \boldsymbol{v} = v\left[\cos\psi\ \boldsymbol{e}_x + \psi_z \boldsymbol{N} + \psi_y \boldsymbol{e}_y\right]\ , \tag{5.6}$$

where the small angles $\theta_x, \theta_y, \theta = \sqrt{\theta_x^2 + \theta_y^2} \sim mc^2/E \ll 1$ define the deviation of the photon wave vector from the centre of PXR reflection, $\psi_z, \psi_y, \psi = \sqrt{\psi_z^2 + \psi_y^2}$ are the deviations of electrons from the incident beam due to multiple scattering and $\boldsymbol{e}_x, \boldsymbol{e}_y$ are the unit vectors (Fig. 5.2). The coherent length with the accuracy $O(\theta, \psi)$ is

$$\frac{1}{L_g} \equiv q = \frac{1}{c}\left[(\boldsymbol{vg}) - \omega(1 - \theta_x - \psi_z)\right] \approx \frac{1}{c}\left[\omega_{\mathrm{B}} - \omega(1 - \theta_x)\right]\ . \tag{5.7}$$

As follows from (5.5), the PXR intensity is maximal at the condition $q = 0$, which defines the position of the PXR spectral line:

$$\omega = \frac{(\boldsymbol{vg})}{1 - \theta_x - \psi_z} \approx \omega_{\mathrm{B}}(1 + \theta_x), \qquad \omega_{\mathrm{B}} = \frac{gc}{\sqrt{2}}\ . \tag{5.8}$$

In this approximation, the radiation frequency is independent of the deviation $\psi$ of the electron velocity from the primary direction, i.e. in the EAD geometry PXR is weakly influenced by multiple scattering. The deviation $\alpha_{\mathrm{B}}$ of the wave vector $\boldsymbol{k}$ from the Bragg condition is:

$$\alpha_{\mathrm{B}} = \frac{2\boldsymbol{kg} + g^2}{k^2} = -\left[\gamma^{-2} + (\theta_y - \psi_y)^2 + (\theta_x - \psi_z)^2\right], \quad \gamma = \frac{E}{mc^2}\ . \tag{5.9}$$

Because the coherence length $L_g$ does not depend on the polarizability, expression (5.5) can be used in (2.36) at the condition $\omega L/c \gg 1$:

$$4L_g^2 \sin^2 L/2L_g = \pi L \delta(q)\ ,$$

which is valid, despite the absorption, for the crystal of thickness $L \gg L_{\mathrm{abs}}$, used in experiments [6, 177]. According to Sect. 1.3, the multiple scattering of electrons can be taken into account by averaging (5.9) over the angles $\psi$:

$$\langle\alpha_{\rm B}\rangle = -\left[\gamma^{-2} + \theta^2 + \theta_s^2(L)\right] , \tag{5.10}$$

where the root-mean-square angle $\theta_s^2(L)$ is determined by (1.25).

Thus, the parameter $\langle\alpha_{\rm B}\rangle$ in the dispersion equation (5.2) is negative in the vicinity of the PXR peak, and the real part of the coefficient $2\gamma_0(\langle\alpha_{\rm B}\rangle + \chi_0)$ in this equation is non-zero due to the condition $\chi_0' < 0$ for the X-ray polarizability. The approximate analytical expressions for roots are then

$$\epsilon_{1s} = -\frac{\chi_0}{2\gamma_0} + \frac{C_s^2|\chi_g|^2}{2\gamma_0(\langle\alpha_{\rm B}\rangle + \chi_0)}, \qquad \epsilon_{2s} = \gamma_g + \sqrt{\gamma_g^2 + \langle\alpha_{\rm B}\rangle + \chi_0} ,$$
$$|\epsilon_{2s}| \sim \sqrt{|\chi_0|} \gg |\epsilon_{1s}| \sim |\chi_0| . \tag{5.11}$$

Due to the condition $|\epsilon_{2s}| \gg |\epsilon_{1s}|$, the principal contribution to the PXR intensity is delivered by the term with $\mu = 1$ in (5.5). In addition, for $2\theta_{\rm B} = 90°, C_1 = 1, C_2 = 0$ (EAD), and the radiation is almost polarized in the direction perpendicular to the plane of vectors $\boldsymbol{N}, \boldsymbol{g}$ ($\sigma$-polarization). This simplifies (5.5) with the accuracy $O(|\chi_0|^2)$:

$$\frac{\partial^2 N_{g\sigma}}{\partial\omega\partial\boldsymbol{n}} = \frac{\alpha\;\omega L\;(\theta_y^2 + \langle\psi_z^2\rangle)|\chi_g|^2\mu_e}{2\pi c^2|\gamma^{-2} + \theta^2 + \theta_s^2(L) - \chi_0'|^2}\delta\left[\frac{g}{\sqrt{2}} - \frac{\omega}{c}(1-\theta_x)\right] ,$$
$$\frac{\partial^2 N_{g\pi}}{\partial\omega\partial\boldsymbol{n}} = \frac{\langle\psi_y^2\rangle}{(\theta_y^2 + \langle\psi_z^2\rangle)}\frac{\partial^2 N_{g\sigma}}{\partial\omega\partial\boldsymbol{n}} . \tag{5.12}$$

Taking into account $\langle\psi_z^2\rangle = \langle\psi_y^2\rangle = \theta_s^2/2$ and the experimental condition $\theta_s^2 \ll |\chi_0'|$, we obtain $N_{g\pi} \ll N_{g\sigma}$.

The spectral intensity of PXR in EAD mode is proportional to $L\omega/c$ for the crystal of any length and can essentially exceed the PXR intensity for conventional Laue and Bragg modes, when the maximal value of $L_{\rm abs}\omega/c$ is determined by the absorption length $L_{\rm abs}(\omega)$ of the radiation in the crystal. The crystal absorption is presented in (5.5) as a factor, depending on the electron coordinate $z_0$. Assuming the beam cross-section $d_e$ and uniform distribution of particles inside a beam, the average of (5.5) over $z_0$ results in the decrease of mean photon numbers per electron, which is taken into account in (5.12) by the factor

$$\mu_e = \frac{1 - {\rm e}^{-2d_e\epsilon_{11}''\omega/c}}{2d_e\epsilon_{11}''\omega/c} < 1 .$$

The absorption of PXR photons in a crystal is less than the absorption of X-rays of the same energy in a homogeneous medium:

$$\epsilon_{11}'' = \chi_0''\left(1 - \frac{|\chi_g|^2}{|\gamma^{-2} + \theta^2 + \theta_s^2(L) - \chi_0'|^2}\right) < \chi_0'' , \tag{5.13}$$

which is a consequence of the known Borrmann effect [25].

According to (1.19) and (5.12), the specific feature of the PXR spectrum inside a single reflection is a discrete sequence of spectral lines. In the case of relativistic electrons, the frequencies of these lines do not depend on the particle energy and correspond to different reciprocal lattice vectors:

$$\omega_{\mathrm{B}}^{(l)} = \frac{cg^{(l)}}{\sqrt{2}} = \frac{\pi c\sqrt{2}}{d_{(hkl)}} l, \qquad l = 1, 2, \dots , \tag{5.14}$$

where $d_{(hkl)}$ is a minimal distance between the reflecting crystallographic planes. In the experiments [5, 44] with the diamond crystal (lattice constant $a = 3.57$ Å), the photon energies in $\{100\}$ and $\{110\}$ reflections are localized at

$$\hbar\omega_{(100)}^{(l)} = \frac{\pi\hbar c\sqrt{2}}{a} l \times 0.63 \times 10^{17}\ \mathrm{keV} \approx 2.46 \times l\ \mathrm{keV}, \qquad l = 4, 8, \dots$$

$$d_{(110)} = \frac{a}{\sqrt{2}}, \qquad \hbar\omega_{(110)}^{(l)} \approx 3.47 \times l\,\mathrm{keV}, \qquad l = 2, 4, 6, \dots . \tag{5.15}$$

To quantitatively fit the theoretical results to the experimental data, the dependence of the spectral distribution and total number of photons in PXR peaks on the angular $\theta_{\mathrm{D}}$ and energy $\delta\omega_{\mathrm{D}}$ resolution of a detector has to be taken into account. Summing over the polarization and integrating over the angle $\theta_x$ in (5.12), the radiation intensity is

$$\frac{\partial^2 N^{(l)}}{\partial\omega\partial\theta_y} = N_0^{(l)} \frac{\theta_y^2 + \theta_s^2}{\omega^{(l)}|\theta_{\mathrm{ph}}^2 + \theta_y^2 + u^2|^2} H\left[\theta_{\mathrm{D}} - |u|\right] ,$$

$$N_0^{(l)} = \frac{\alpha\, \omega^{(l)} L\, |\chi_g|^2 \mu_e}{2\pi c}, \qquad \theta_{\mathrm{ph}}^2 = \gamma^{-2} + \theta_s^2 - \chi_0', \qquad u = \frac{\omega}{\omega^{(l)}} - 1, \tag{5.16}$$

where $H[z]$ is a stepwise Heaviside function. The natural spectrum of the PXR line near the frequency $\omega^{(l)}$ is then

$$\frac{\partial N^{(l)}}{\partial\omega} = \frac{N_0^{(l)}}{\omega^{(l)}} \left\{ \frac{|u^2 + \theta_{\mathrm{ph}}^2| + \theta_s^2}{|u^2 + \theta_{\mathrm{ph}}^2|^{3/2}} \arctan \frac{\theta_{\mathrm{D}}}{\sqrt{|u^2 + \theta_{\mathrm{ph}}^2|}} \right.$$
$$\left. + \frac{\theta_{\mathrm{D}}(\theta_s^2 - |u^2 + \theta_{\mathrm{ph}}^2|)}{|u^2 + \theta_{\mathrm{ph}}^2|(\theta_{\mathrm{D}}^2 + |u^2 + \theta_{\mathrm{ph}}^2|)} \right\} H\left[\theta_{\mathrm{D}} - |u|\right] \tag{5.17}$$

and depends on the ratio between the kinematic parameter $\theta_{\mathrm{ph}}$ and two instrumental parameters $\theta_{\mathrm{D}}$ and $\delta\omega_{\mathrm{D}}$. The value $\theta_{\mathrm{D}} \sim S_{\mathrm{D}}/R$ ($S_{\mathrm{D}}$ is the area of the detector slit) depends on the distance $R$ between the crystal and the detector. According to classification [40], the 'far case' ($\theta_{\mathrm{D}} < \delta\omega_{\mathrm{D}}/\omega^{(l)}$) or 'near case' ($\theta_{\mathrm{D}} > \delta\omega_{\mathrm{D}}/\omega^{(l)}$) mode can be realized in the experiment.

Reflections from the planes $\{100\}$ and $\{110\}$ with other indices are equal to zero; $L_{\mathrm{D}}$ is the total distance between the crystal and the detector; $\eta$ is the detector efficiency; the value $\mu_e = \left[1 - \exp(d/L_{\mathrm{abs}}^{(c)})\right] L_{\mathrm{abs}}^{(c)}/d$ is fulfilled as a result of averaging on the electron coordinates $z_0$ in (5.5).

**Table 5.2.** Experimental parameters for spectra simulation

| Plane index | $\omega_B^{(l)}$ (keV) | $\omega_{exp}^{(l)}$ (keV) | $\lvert\chi_0\rvert$ $\times 10^6$ | $\lvert\chi_g\rvert^2$ $\times 10^{13}$ | $\theta_{ph}^2 \times 10^6$ (rad$^2$) | $L_{abs}^{(c)}$ (cm) diamond | $L_{abs}^{(w)}$ (cm) window | $L_{abs}^{(a)}$ (cm) air | $\eta$% | $\mu_e$ |
|---|---|---|---|---|---|---|---|---|---|---|
| (400) | 9.82 | 9.7 ± 0.15 | 15.31 | 120.0 | 18.58 | 0.12 | 0.148 | 136 | 78 | 0.73 |
| (800) | 19.65 | 19.7 ± 0.23 | 3.83 | 1.0 | 6.77 | 0.63 | 1.003 | 1,058 | 15 | 0.94 |
| (220) | 6.95 | 6.8 ± 0.15 | 30.62 | 883.0 | 33.89 | 0.04 | 0.045 | 52 | 94 | 0.24 |
| (440) | 13.84 | 13.8 ± 0.15 | 7.66 | 14.8 | 10.60 | 0.29 | 0.342 | 388 | 27 | 0.87 |
| (660) | 20.84 | 20.7 ± 0.20 | 3.40 | 0.6 | 6.35 | 0.75 | 1.026 | 1,262 | 15 | 0.98 |

| $L$ (cm) | $L_1$ (cm) | $L_2$ (cm) | $L_D$ (cm) | $a$ (cm) | $d$ (cm) | $E$ (MeV) | $L_R$ (cm) | $\theta_D \times 10^2$ (rad) |
|---|---|---|---|---|---|---|---|---|
| 0.1 ± 0.02 | 0.6 | 155 | 187 | 2.5 | 0.08 | 900 | 14.8 | 1.30 |
| | [44] | [44, 78] | [44, 78] | [44, 78] | | | | [44, 78] |
| – | 0.005 | 72 | 104 | 0.65 | – | – | – | 0.63 |
| | [30, 78] | [30] | [44, 78] | [30] | | | | [30] |

In experiments [5, 30, 44, 78], the following conditions were used (Table 5.2):

$$\theta_{ph} \ll \theta_D \ll \delta\omega_D/\omega^{(l)} . \tag{5.18}$$

The real shape of the PXR line is defined by the instrumental function of the detector, and data fitting is based on the integral photon number $N_{th}^{(l)}$ in the peak, which follows from (5.17):

$$N_{th}^{(l)} = \frac{\pi}{2} N_0^{(l)} \left\{ \ln(D_l^2 + 1) - \frac{D_l^2}{D_l^2 + 1} \right\}, \quad D_l = \frac{\theta_D}{|\theta_{ph}|} \gg 1 . \tag{5.19}$$

There are also additional contributions to the observed number of photons $N_D^{(l)}$, from coherent bremsstrahlung and diffracted bremsstrahlung (DBS). Most of the bremsstrahlung photons from the relativistic charged particle moving in a medium are concentrated in a narrow cone around the particle velocity [15]. These photons can be reflected by crystallographic planes and diffract in the directions of PXR reflections [35].

The DBS contribution to the reflection intensity can be estimated if PXR is considered as a result of pseudophoton diffraction (Sect. 1.2). The number of photons in the PXR peak, corresponding to the reciprocal lattice vector $\boldsymbol{g}$, is

$$N_{PXR}(\boldsymbol{g}) \approx n(\omega) R(\boldsymbol{g}) \delta\omega_D , \tag{5.20}$$

where $n(\omega) \approx (\alpha/2\pi\omega) \ln(E/\hbar\omega)$ is the spectral density of pseudophotons and $R(\boldsymbol{g})$ is the probability of the reflection of photons with frequency $\omega$ from crystallographic planes. The number of DBS photons is

$$N_{DBS}(\boldsymbol{g}) \approx n_{BS}(\omega) R(\boldsymbol{g}) \delta\omega_D . \tag{5.21}$$

The spectral density of bremsstrahlung quanta, emitted on the length $L_{\rm ext}$ inside a crystal with atoms of charge $Z$ and volume $\rho_0$, is [35]

$$n_{\rm BS}(\omega) \approx \frac{16}{3} Z^2 \alpha r_e^2 \frac{\rho_0 L_{\rm ext}}{\omega} \ln\left(\frac{E^2}{\hbar m c^2 \omega}\right) . \tag{5.22}$$

Here $r_e$ is the electromagnetic radius of an electron, and $L_{\rm ext}$ is inversely proportional to the crystal polarizability:

$$L_{\rm ext} = \frac{1}{\omega|\chi_0|}, \qquad |\chi_0| \approx \frac{4\pi Z r_e c^2}{\omega^2} \rho_0 . \tag{5.23}$$

From (5.20)–(5.21), the estimate for the DBS and PXR ratio is straightforward:

$$\xi_1 = \frac{N_{\rm DBS}}{N_{\rm PXR}} \approx \frac{4\omega}{3c} Z r_e \left[1 + \frac{\ln\gamma}{\ln(E/\hbar\omega)}\right] . \tag{5.24}$$

For the considered experiments, $\omega \approx 10$ keV, $Z = 6, E = 900$ MeV, the parameter $\xi_1 \approx 3\times10^{-3}$; thus the DBS contribution to the total reflection intensity can be neglected. The relative contribution of coherent bremsstrahlung to the total intensity is also inessential and derived from (3.23):

$$\xi = \frac{N_{\rm CBS}}{N_{\rm PXR}} \approx \left(\frac{m^2c^4}{E\hbar g}\right)^2 \frac{1}{16\sin^4\theta_{\rm B}} \approx 10^{-4} . \tag{5.25}$$

Figure 5.3 compares the values $N_{\rm exp}^{(l)}$ calculated from the spectra in Fig. 1.7 by taking into account the detector background correction and its intrinsic

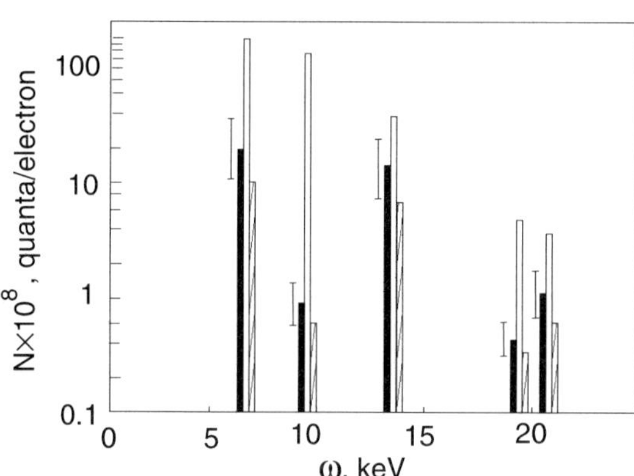


**Fig. 5.3.** The diagram of the experimental and theoretical values for PXR quanta emitted in various reflections. *Solid area*: the number of quanta counted by the detector $N_{\rm exp}^{(l)}$; *unshaded area*: theoretical number $N_{\rm th}^{(l)}$ of PXR quanta generated in the crystal; *shaded area*: the values $N_{\rm D}^{(l)}$ calculated by taking into account the corrections on the photon absorption between the crystal and the detector and on the detector efficiency in accordance with Table 5.2

spectral function in the Gaussian form. Besides, we take into account the PXR absorption at the accelerator exit, the detector input windows and in the air between the crystal and the detector. Various radiation harmonics have different absorption lengths $L_{\rm abs}^{(l)}$ in these media and, as a result, the correlation between the values $N_{\rm exp}^{(l)}$ for different peaks is essentially changed in comparison with the corresponding values $N_{\rm th}^{(l)}$. All parameters required for calculation are shown in Table 5.2. The number of quanta $N_{\rm D}^{(l)}$ calculated by taking into account these corrections fits the experimental results quite well.

## 5.3 Investigation of the Production Mechanism of PXR

By now, numerous experimental and theoretical works have been published, investigating the production mechanism of PXR. In this section, we consider some of them, where the qualitative properties of PXR emission and detection have been studied.

As follows from (1.19), the formation of the series of harmonics at a fixed detector and crystal position is one of the important unique features of PXR. The detailed investigation of these series was carried out in [39, 49]. In the experiments [39], the intensity of PXR reflections from 855 MeV electrons transmitting through the silicon crystal was measured. By rotating the crystal, the intensity of harmonics, corresponding to different crystallographic planes but at fixed angle $\theta_{\rm B} = 22.5°$, was recorded.

The typical series of the PXR reflection from the set of {111} planes is shown in Fig. 5.4. The position of the peaks is well explained by (1.19) if the lattice constant $a = 5.43$ Å is used:

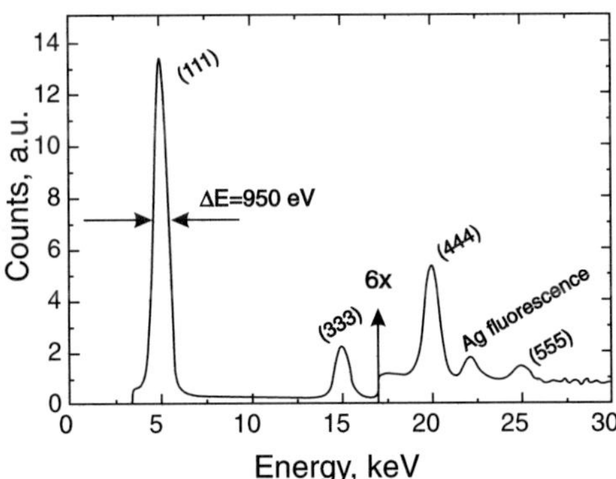


**Fig. 5.4.** The energy spectrum of PXR for the beam energy $E = 855$ MeV and the Bragg angle $\theta_{\rm B} = 22.5°$

$$\hbar\omega^{(l)}_{(111)} = \frac{\pi\hbar c\sqrt{3}}{a\sin\theta_{\rm B}} l \times 0.63 \times 10^{17}\ \text{keV} \approx 5.12 \times l\ \text{keV}\ ,$$
$$l = 1, 3, 4, 5\ldots\ . \tag{5.26}$$

The theoretical interpretation of PXR spectra, which uses an averaging of coherent length [48] for taking into account the multiple electron scattering, failed to explain the experiment [39]. This failure can be understood from general analysis of multiple scattering of electrons, considered in Sect. 1.3 on the basis of the classical theory. In the general case of thick crystals, the total PXR output is obtained from the averaging of intensity over all electron trajectories, defined by the kinetic equation [29]. However, this task can be essentially simplified, and an analytical solution can be obtained in two limiting cases, which are often realized in experiments. The first case [48] corresponds to the limitation of the crystal thickness:

$$L < L_0, \qquad \theta_{\rm s}^2(L_0) = \gamma^{-2} + |\chi_0|\ , \tag{5.27}$$

i.e. multiple scattering does not change the radiation coherence over entire particle trajectory in the medium. This approximation for multiple scattering is called coherent model. Using (1.25) for the average angle of multiple scattering and explicit expression for polarizability, the parameter $L_0(\omega, E)$ is

$$L_0(\omega, E) = L_R\left[\frac{mc^2}{E_{\rm s}^2} + \frac{\omega_0^2}{\omega^2}\frac{E^2}{E_{\rm s}^2}\right]\ . \tag{5.28}$$

On the other hand, according to Sect. 1.3, the approximate averaging over the trajectories as well as over mosaicity of crystal can also be done for thick crystals if the thickness and energy (or angular) resolution of the detector satisfies the conditions

$$L > L_0, \qquad \frac{\delta\omega_{\rm D}}{\omega_{\rm B}} > \theta_{\rm s}^2(L)\ , \tag{5.29}$$

which are applicable to [39].

The crystal in this case can be considered as the aggregate of blocks, each with the thickness $L_i \sim L_0$. The radiation from these blocks is formed incoherently; however, due to low energy resolution of the detector, the PXR reflections from separate blocks are located within the single peak. Thus, the total number of photons in the peak is proportional either to the total crystal thickness $L = \sum L_i$ or to the absorption length $L_{\rm abs}$ if $L > L_{\rm abs}$. The procedure of averaging over the blocks in calculation of the PXR intensity is analogous to the one used in X-ray diffractometry [46, 69], where the scattering amplitude from crystallographic planes within a single block of a mosaic crystal is calculated by the kinematic theory, and the integral intensity is proportional to $L$ or $L_{\rm abs}$.

In [41, 46], the incoherent model has been developed, useful for practical applications, for example for [39]. The applicability of this model is restricted by (5.29). To use it, the factor $f(\boldsymbol{N},\boldsymbol{v},\boldsymbol{k})$ is implemented in the kinematic equation (2.44), which approximately takes into account the absorption and the deviation of electron trajectory from the primary direction. The following expression results for the angular distribution of photons in the PXR reflection:

$$\frac{\partial^2 N_{\boldsymbol{g}}}{\partial\theta_x\partial\theta_y} = \frac{\alpha}{4\pi c}\sum_{l=1}^{\infty}\omega_{\mathrm{B}}^{(l)} L_{\mathrm{abs}}\frac{\left|\chi_g(\omega_{\mathrm{B}}^{(l)})\right|^2}{\sin^2\theta_{\mathrm{B}}}\frac{\theta_x^2\cos^2 2\theta_{\mathrm{B}}+\theta_y^2}{(\theta_x^2+\theta_y^2+\gamma^{-2}+|\chi_0|)^2} f(\boldsymbol{N},\boldsymbol{v},\boldsymbol{k}) \,,$$

$$f(\boldsymbol{N},\boldsymbol{v},\boldsymbol{k}) = \left|\frac{\cos(2\theta_{\mathrm{B}}-\phi)}{\cos\phi}\right|\left(1-\mathrm{e}^{-L/L_{\mathrm{abs}}|\cos\phi|}\right), \qquad \phi = \widehat{\boldsymbol{v}\boldsymbol{N}} \,. \quad (5.30)$$

For simulation of the detected intensity in [39], the convolution of (5.30) with the Gaussian angular distribution $\sigma_{\mathrm{S}}$ for electrons due to multiple scattering has been used:

$$\sigma_{\mathrm{S}} = \frac{13.6\ \mathrm{MeV}}{E}\sqrt{\frac{L}{L_R}}\,[1+0.038\ \ln(L/L_R)] \ .$$

Table 5.3 demonstrates a very good agreement between the theoretical and experimental values of the photon flux per steradian and single electron calculated and measured in [39] by means of the method described above. The real values of $L_{\mathrm{abs}}$ are compared with those obtained from the fit of the experimental data.

**Table 5.3.** Measured and calculated photon flux of PXR for the (111), (220) and (224) reflections at the maximum of the angular distribution [40]

| Plane | Energy (eV) | Photon Flux Measured | $\mathrm{d}N/\mathrm{d}\Omega$, e$^-$ sr Calculated | $L_a^{\mathrm{th}}$ µm [45] | $L_a^{\mathrm{exp}}$ µm |
|---|---|---|---|---|---|
| (111) | 5,166 | $(4.5 \pm 0.5)\times 10^{-3}$ | $4.3\times 10^{-3}$ | 20.2 | – |
| (220) | 8,332 | $(6.5 \pm 0.8)\times 10^{-3}$ | $6.3\times 10^{-3}$ | 84.9 | $91 \pm 5$ |
| (224) | 14,630 | $(5.4 \pm 0.5)\times 10^{-3}$ | $5.2\times 10^{-3}$ | – | – |
| (440) | – | – | – | 657.8 | $649 \pm 80$ |

The effectiveness of the incoherent model for multiple scattering in thick crystals is confirmed in [24, 45], where the dependence of PXR on the crystal thickness and electron energy has been investigated. In these experiments, the integral PXR output from {111} and {110} planes in a $0.2\ \mathrm{mm} < L < 5$ mm thick Si crystal and electron energy $200\,\mathrm{MeV} < E < 1100\,\mathrm{MeV}$ has been measured. The typical PXR spectrum from [24, 45] is shown in Fig. 5.5. For the silicon crystal, the radiation length $L_R = 5.1\,\mathrm{cm}$ and the plasma frequency $\omega_0 = 1.2\times 10^{15}\ \mathrm{s}^{-1}$. For the electron energy $E = 500\,\mathrm{MeV}$ and photon energy $\hbar\omega = 10$ keV, the value $L_0 \simeq 0.1\,\mathrm{mm}$; i.e. in [24, 45] condition (5.29) for

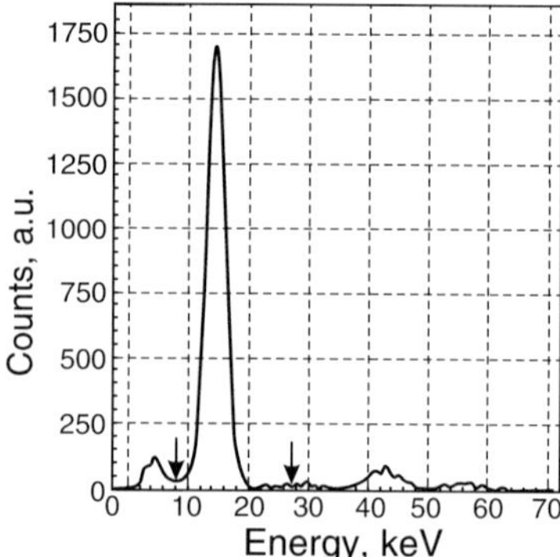


**Fig. 5.5.** A typical spectrum of PXR for the electron energy 1,100 MeV from [45]. The upper and lower discrimination levels for counting the PXR photons belonging to the (111) reflection are shown by the arrows

applicability of the incoherent model has been satisfied. Figures 5.6 and 5.7 from the papers [24, 45] show that the dependence of the PXR intensity on the parameters $L$ and $E$ can be simulated quite reasonably in the framework of the 'incoherent model' if conditions (5.29) are fulfilled.

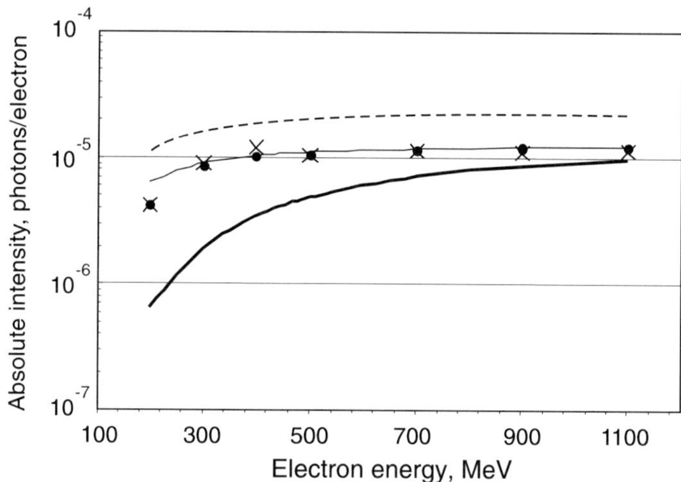


**Fig. 5.6.** The electron energy dependence of the PXR photon numbers accepted by the detector solid angle $9.12 \times 10^{-5}$ sr for $\theta = \theta_{\mathrm{B}}$. The experimental values are plotted by solid circles. The solid and dashed curves are the theoretical predictions of the coherent and incoherent model, respectively. The thin line shows the simulation of the multiple scattering by means of a more rigorous method [63]

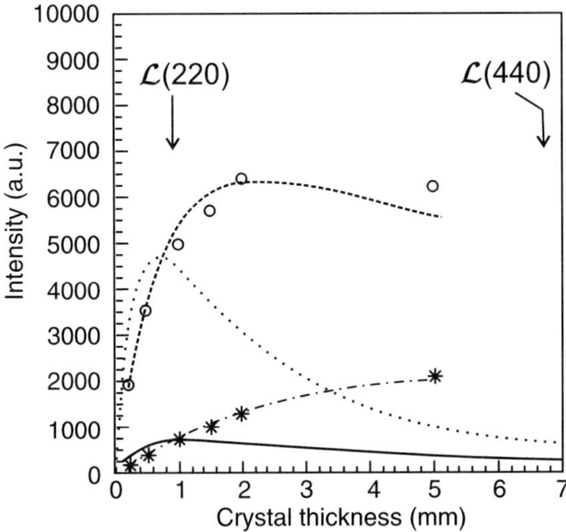


**Fig. 5.7.** Crystal thickness dependence of the PXR intensity. The *circles* are for the (220) reflection and the *asterisks* are for the (440) reflection. The *solid* and *dashed lines* are predictions of the coherent model for the (220) and (440) reflections. The *dotted* and *dot-dashed lines* are the simulation by the incoherent model for the same reflections. The *arrows* show the X-ray absorption length in the crystal for the (220) and (440) reflections, respectively

A very important role for understanding the microscopic nature of PXR is played by the experiment [49], where low-energy electrons ($E = 90\,\mathrm{MeV}$) were used for monitoring intensity damping in the peak sequence of the photon energy spectrum depending on the harmonics number $l$. The results for the silicon target are well explained by the incoherent model. For a pyrolytic graphite sample, the maximal up to date number of harmonics $l_{\max} = 8$ for the (111) reflection has been detected. The surprise was the underestimate of the theoretical intensity $I_l$ for higher harmonics [49], for example $I_8^{\exp}/I_8^{\mathrm{th}} \simeq 200$. The interpretation failure was explained [49] by incorrect mosaicity consideration in the PXR emission process. The discrepancy between the theoretical predictions and experimental measurements [49] for high harmonics was also discussed in [79]. Diffraction of the bremsstrahlung quanta and the loss of some emitted quanta due to the pulse shortness were assumed as possible reasons for this discrepancy.

The considerable gain of intensity in higher harmonics, however, may also be related to the contribution of coherent bremsstrahlung to the peaks with a higher $l$. The resonant frequencies of PXR and CBS (Sect. 3.3) are defined by the same kinematic factors, therefore the intensity (3.19) depends on the amplitudes of both processes. For the mosaic graphite crystal in [49], condition (5.29) is fulfilled, and (3.19) can be simplified in the framework of the incoherent model, using a kinematic approximation (5.29). Thus, the angular distribution of the photons at fixed $\theta_{\mathrm{B}}$ in the spectral peak with index $l$ is given by

$$\frac{\partial^2 N_l}{\partial\theta_x\partial\theta_y} = \frac{\alpha}{4\pi c}\omega_l L_{\mathrm{abs}}(\omega_l) f(\boldsymbol{N},\boldsymbol{v},\boldsymbol{k})\frac{(\theta_x^2\cos^2 2\theta_{\mathrm{B}}+\theta_y^2)}{\sin^2\theta_{\mathrm{B}}}$$
$$\times\left|\frac{\chi_{g_l}(\omega_l)}{(\theta_x^2+\theta_y^2+\gamma^{-2}+|\chi_0|)} - \frac{4\pi\alpha[Z-F(g_l)]\mathrm{e}^{-W(g_l)}}{\sin\theta_{\mathrm{B}} l^3}\left(\frac{d}{a}\right)^3\right|^2 ,$$
$$g_l = \frac{2\pi}{d}l, \qquad \omega_l = c\frac{g_l}{2\sin\theta_{\mathrm{B}}} . \tag{5.31}$$

Here $F(g_l)$ is the atomic scattering factor and $\mathrm{e}^{-W(g_l)}$ is the Debye–Waller factor; $d$ is the distance between the planes for the considered reflection and $a$ is the lattice constant. Taking into account $\chi_{g_l}(\omega_l) \sim F(g_l)\mathrm{e}^{-W(g_l)}$, it is evident that the first harmonics are mainly contributed by PXR for $Z \simeq F(g_l)$ and $E = 90\,\mathrm{MeV}$, as follows from (5.31). However, due to exponential damping of the atomic factor for a higher $l$, the relative contribution of CBS (the second term in (5.31)) becomes principal. The less the atomic charge of the crystal, the more pronounced the effect. It is important that the PXR contribution to the intensity reduces as $\sim \theta^{-2}$, whereas the CBS contribution increases as $\sim \theta^2$. Therefore, the ratio between the numbers of detected PXR and CBS quanta strongly depends on the angular resolution of the detector.

Figure 5.8 shows the results of CBS consideration in the PXR spectrum for the experiment [49]. The contributions of PXR and CBS are shown separately for several harmonics from [49]. All the theoretical curves are simulated by integration over the angles in (5.31) within the intervals $\theta_{\mathrm{D}} = \gamma^{-1}$, $10\gamma^{-1}$ rad

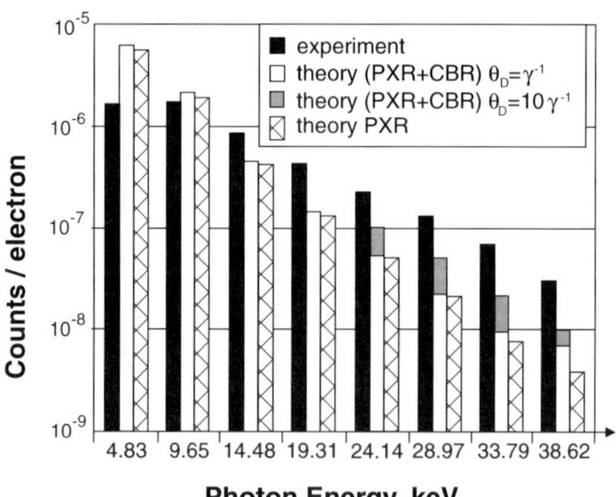


**Fig. 5.8.** Spectral harmonics of the radiation from the electron of $E = 90\,\mathrm{MeV}$ in the graphite mosaic crystal for the (002) series with $\theta_{\mathrm{B}} = 22.5°$ [49] and their theoretical simulations

and using $F(g_l)$ from [47, 77]. The above results show that a more detailed study is necessary for accurate explanation of the experiment.

The dependence of the PXR the peak frequency on the crystal and detector orientation with respect to the electron beam is important for applications. For arbitrary electron energy, the relation between these parameters is defined by (4.16):

$$\omega_l = \frac{(\boldsymbol{v}\boldsymbol{g}_l)}{1 - (\boldsymbol{v}\boldsymbol{k})/ck} . \tag{5.32}$$

In general, the peak frequency of spectral series depends on both the angle $\theta_{\rm B}$ between the velocity and reflecting planes $\widehat{\boldsymbol{v}\boldsymbol{g}_l} = \pi/2 - \theta_{\rm B}$ and the angle $\theta_0$ between the velocity and detector (Fig. 2.3). In experiments, where the angular dependence of $\omega_l$ was investigated, the relativistic electrons were used: $(1 - v^2) = \gamma^{-2} \ll 1$. Then according to (2.38), the wave vector of the emitted photon is located near the vector $\boldsymbol{k}_{\rm B}$, $\boldsymbol{k} = \boldsymbol{k}_{\rm B} + \boldsymbol{u}$, $u/k_{\rm B} \sim \gamma^{-1} \ll 1$. Because $\widehat{\boldsymbol{v}\boldsymbol{k}_{\rm B}} = 2\theta_{\rm B}$, the peak frequency depends on the single angle only:

$$\omega_l \approx \frac{g_l c}{2 \sin \theta_{\rm B}} (1 + \theta_x \cot \theta_{\rm B}) , \tag{5.33}$$

where $\theta_x \sim \gamma^{-1}$ defines the projection of the vector $\boldsymbol{k} - \boldsymbol{k}_{\rm B}$ onto the plane of the vectors $\boldsymbol{v}$ and $\boldsymbol{k}_{\rm B}$ (Fig. 2.3).

Figure 5.9 shows the experimental measurements of the PXR photon energy as a function of angles for different scan types. In [53], the intensity of the (111) reflection from a diamond crystal and an electron of energy $E = 8.3\,\text{MeV}$ were measured for simultaneous rotation of the crystal around $\boldsymbol{v}$ and keeping the detector on the maximal reflection intensity. This scan is described by (5.33) with $\theta_x = 0$ and is, in fact, represents the DuMond diagram for this reflection. The measurements in [66] have been carried out with electrons of

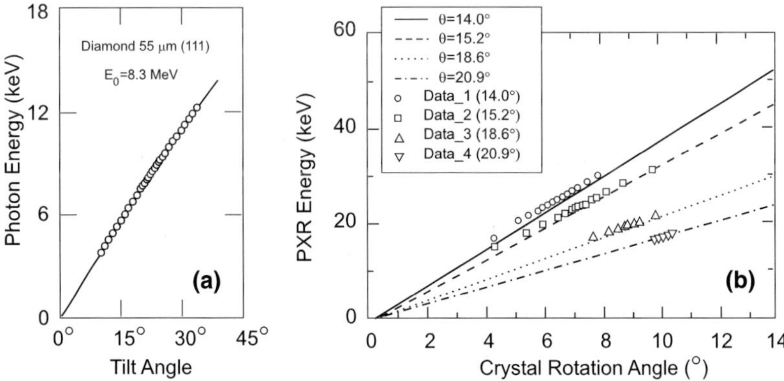


**Fig. 5.9.** Angular dependence of the PXR photon energy: (**a**) the value $\theta_{\rm B}$ was varied with $\theta_0 = 2\theta_{\rm B}$ [53]; (**b**) the value $\theta_0 = 2\theta_{\rm B}$ was fixed and the crystal rotated by the angle $\theta_x$ near the value $\theta_{\rm B}$ [67]

$E = 45\,\text{MeV}$ and a Si crystal for the (220) reflection. The detector position was fixed at $\theta_0 = 2\theta_B$, and the crystal was rotated by a small angle $\theta_x$ with respect to $\theta_B$ to remain on the peak. For this scan, the energies of the detected photons are on the straight lines, corresponding to (5.33) at a fixed $\theta_B$.

Thus, the spectral and intensity experimental PXR features from relativistic electrons are well explained by the kinematic theory with correction for multiple scattering, detector instrumental function and coherent bremsstrahlung.

## 5.4 Angular Distribution and Polarization of PXR

The fine structure of the PXR angular distribution and polarization of photons in the reflection follows directly from the kinematic PXR theory [46, 48]. The first experiments (see Sect. 1.4) already qualitatively confirmed the asymmetry of the intensity in angular distribution with respect to the centre of reflection. The detailed experimental study of the PXR reflection fine structure has been performed in [73].

Figure 5.10 shows the experimental angular distribution of PXR in the (111) reflection for the Laue geometry from a 17 μm-thick Si crystal and simulated on the basis of the kinematic theory of X-ray intensity [73]. As shown in the experimental sketch (Fig. 5.10a), the angular distribution was scanned at a fixed observation angle $\theta_0 = 0.306$ rad with respect to the beam velocity by rotating the crystal by an angle $\Phi$. The angular variables for intensity distribution are expressed via $\Phi$ using (2.51)–(2.54), and after integration over the frequencies the PXR intensity is (absorption included)

$$\frac{\partial N}{\partial \Phi} = \frac{\alpha}{2\pi}\frac{J}{e}\sum_{l=1}^{\infty}\left|\chi_{g_l}(\omega_B^{(l)})\right|^2 \frac{\omega_B^{(l)} L_{\text{abs}}(1 - \mathrm{e}^{-L/L_{\text{abs}}})}{c\sin^2\theta_B}$$
$$\times \frac{(\Delta\Phi)^2\cos^2 2\theta_B}{\left[(\Delta\Phi)^2 + \gamma^{-2} + |\chi_0(\omega_B^{(l)})|\right]^2}\,,$$

$$\omega_B^{(l)} = \frac{\pi c\sqrt{3}}{a\sin\theta_B}l, \quad a = 5.43\,\text{Å}, \quad \theta_B = \frac{\theta_0 + \Delta\Phi}{2}, \quad \Delta\Phi = \Phi - \Phi_0\,. \tag{5.34}$$

The centre of the PXR reflection is positioned at $\Phi_0 = 0.153$ rad, whereas the maxima are located at $\Delta\Phi_\pm = \pm\sqrt{\gamma^{-2} + |\chi_0|}$. In contrast to the curve in Fig. 2.5, there is an asymmetry in the angular distribution of intensity, which is explained by different values of the factor $\cos^2 2\theta_B/\sin^2\theta_B$ at points $\Delta\Phi_\pm$ for low-energy electrons $\gamma^{-2} \gg |\chi_0|$. The larger the electron energy $E$, the smaller the asymmetry. The existence of a close to zero intensity minimum at $\Phi = \Phi_0$ means that multiple electron scattering was negligible for the thickness of the crystal used. The simulation [73] using the formula analogous to (5.34), fits well the experimental data (Fig. 5.10).

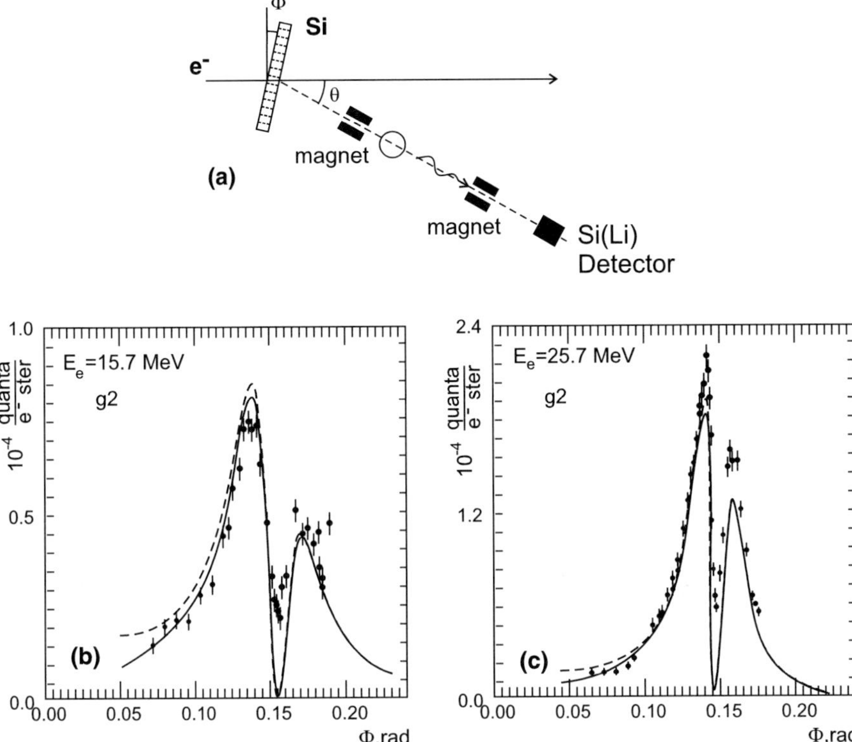


**Fig. 5.10.** Angular distributions of the photons in the PXR reflection in accordance with [73] (Laue case): (**a**) the experimental set-up; (**b**) experimental data for the photon intensity as a function of $\Phi$ from the electrons with energy $E = 15.7\,\text{MeV}$ (points), theoretical simulation without taking into account the absorption of the photons in a crystal (dashed line) and with absorption (solid line); (**c**) the same for the electron energy $E = 25.7\,\text{MeV}$

The PXR angular distribution of electrons of $E = 100\,\text{MeV}$ in the Bragg geometry was measured in [55]. The (111) reflection of a silicon crystal was used, and a detector was placed at an angle $\theta_0 = 144°$ to the electron beam.

The experimental data and simulations are depicted in Fig. 5.11. There is no intensity minimum at the centre of reflection due to the increase of the multiple scattering angle, which corresponds to the theory of Sect. 1.3 for thick crystals. The theoretical simulations in [55] demonstrate the dynamical diffraction effects for the perfect crystals, which lead to the increase of the absorption length, according to (5.13).

The experiments [51] with the (220) PXR reflection from the 20-μm Si crystal, $\theta_B = 45°$ and electrons with $E = 230\,\text{MeV}$, present for the first time a map of the angular distribution using a special imaging detector. According to the PXR theory for this geometry, the pattern is almost completely

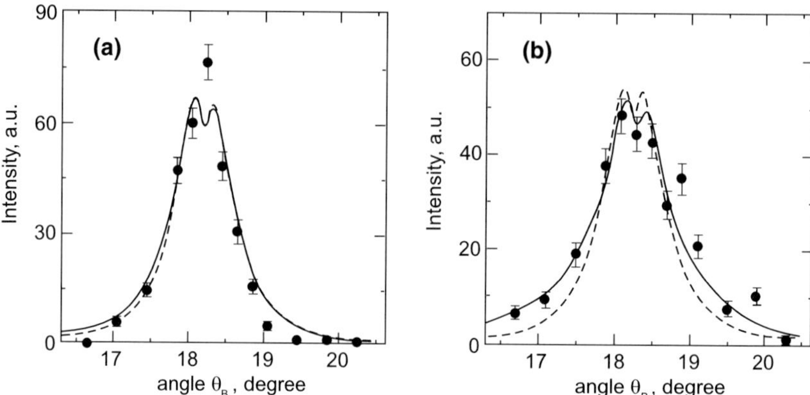


**Fig. 5.11.** Angular distributions of the photons in the PXR reflection according to [55] (Bragg case): (**a**) experimental data for the photon intensity as a function of $\theta_B$ from the electrons with energy $E = 100\,\mathrm{MeV}$ and the crystal thickness $L = 250\,\mu\mathrm{m}$ (points); theoretical simulation with the absorption of the photons in a crystal (dashed line) and without absorption (solid line); (**b**) the same for $L = 3,000\,\mu\mathrm{m}$

polarized along the direction $\theta_y$ perpendicular to the plane of the vectors $\boldsymbol{v}$ and $\boldsymbol{g}$. Moreover, the analysis of the contribution from different harmonics into PXR angular distribution is carried out (Fig. 5.12). In [39], the analysis of contributions by other than PXR radiative mechanisms into reflections of different orders was performed. The experimental results (Fig. 5.13) demonstrate an increased role of the non-PXR radiations in the reflections, which agrees with the estimate (5.31) for a graphite crystal.

Concerning the polarization of PXR, the investigation of anisotropy effects for the scattering of X-rays in a medium is not really done today, as it was for optical diapason [38], for example. The tensor of the dielectric permittivity for X-rays is, in contrast to the optic range, isotropic for almost the entire range of frequencies, except for the narrow region near characteristic atomic lines and nuclear resonant transitions (Appendix A.1). Preferential vectors of polarization (eigenvectors) $\boldsymbol{e}_s \perp \boldsymbol{k}$, $s = 1, 2$, for X-rays in crystals are caused by the interference of the waves reflected from the crystallographic planes. Therefore, they are defined by the experimental geometry (the Bragg angle $\theta_B$ and reciprocal lattice vectors $\boldsymbol{g}$), but not by the atomic characteristics. In the most usual case of two-beam diffraction, the eigenvectors are $\pi$-polarized ($\boldsymbol{e}_\pi$ is in the plane of the vectors $\boldsymbol{k}$ and $\boldsymbol{g}$) and $\sigma$-polarized ($\boldsymbol{e}_\sigma$ is perpendicular to the plane of $\boldsymbol{k}$ and $\boldsymbol{g}$) [56]. The diffraction theory for both waves differs by the polarization factor $C_s$ in the Fourier component of the X-ray polarizability:

$$\chi_g^s = \chi_g C_s , \qquad C_\sigma = 1 , \qquad C_\pi = \cos 2\theta_B . \tag{5.35}$$

For conventional X-ray diffraction, the radiation reflected from the crystallographic planes consists of an incoherent mixture of waves with $\sigma$- and

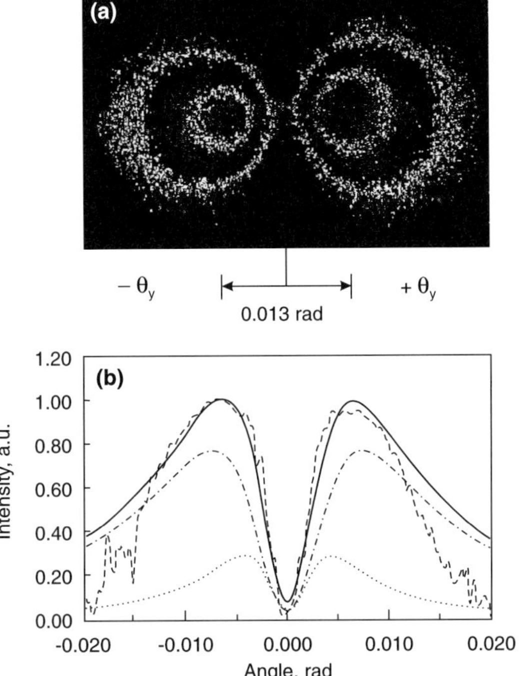


**Fig. 5.12.** Some results from [51]: (**a**) image of the PXR angular distribution from the (220) reflection, $\theta_{\mathrm{B}} = 45°$; (**b**) vertical scan of intensity as a function of angle $\theta_y$ (points), harmonics $l = 1$ (*dot-dashed line*), $l = 2$ (*dashed line*), contribution by both harmonics (*solid line*)

$\pi$-polarizations. In the case of the Bragg diffraction, the contribution of $\pi$-polarization is

$$P_\pi = \frac{1}{1 + \cos^2 2\theta_{\mathrm{B}}}, \qquad 0.5 < P_\pi < 1 , \tag{5.36}$$

whereas in the Laue diffraction, the parameter $P_\pi$ varies between the same limits, but depends on both $\theta_{\mathrm{B}}$ and the crystal thickness [25].

The polarization properties of PXR from relativistic particles are similar to those of X-rays in conventional diffraction. According to the two-beam approximation for PXR (Sect. 1.3), the intensity of radiation in the PXR reflection is composed of two incoherent photon fluxes with $\sigma$- and $\pi$-polarizations. Therefore, the degree of PXR polarization for the reciprocal vector $\boldsymbol{g}$ is introduced as a relative intensity of $\pi$-polarized photons:

$$P_\pi(\boldsymbol{g}) = \frac{\Delta N_\pi}{\Delta N_\pi + \Delta N_\sigma} ,$$

$$\Delta N_s = \int\int_{\Delta\Omega_{\mathrm{D}}} \mathrm{d}\theta_x \mathrm{d}\theta_y \int_{\Delta\omega_{\mathrm{D}}} \frac{\partial^2 N_{\boldsymbol{g}}^{(s)}}{\partial\Omega\partial\omega}; \qquad s = \sigma, \pi , \tag{5.37}$$

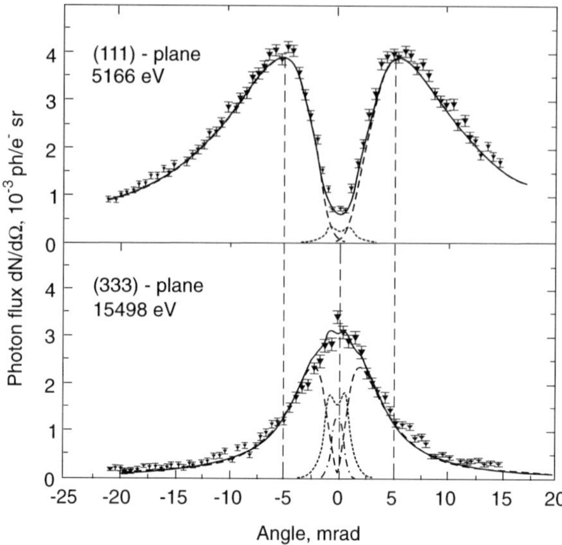


**Fig. 5.13.** Angular $\theta_x$-distribution of PXR at the Bragg angle $\theta_{\mathrm{B}} = 22.5°$ [39]. The *top* plot shows the (111)-distribution and the *bottom* plot the (333)-distribution. The experimental data are represented by circles, and the curves show the contribution by PXR (*broken line*), transition radiation (*point*) and bremsstrahlung (*dash-dotted*)

where the integration over the angular and frequency resolution of the detector is performed.

In kinematic approximation for PXR (2.42), where $\varphi = 0$ for $\sigma$-polarization and $\varphi = \pi/2$ for $\pi$-polarization, and for real detectors ($\Delta\omega_{\mathrm{D}}/\omega > \gamma^{-1}$), the integrals in (5.37) are cancelled and PXR polarization can be calculated from (5.36), like in the case of X-ray diffraction.

When the incoherent model for crystal mosaicity and multiple scattering is adopted, formula (2.51) for the PXR spectral density may be used, and PXR polarization is calculated by (5.36). For PXR in perfect crystals, however, the calculation of $P_\pi(\boldsymbol{g})$ with dynamical effects requires the numerical integration, and the result depends on the crystal thickness. In the target with $L > L_{\mathrm{ext}}$, the polarization depends on $\theta_{\mathrm{B}}$ only. The behaviour of the $P_\pi(\theta_{\mathrm{B}})$ dependence is illustrated in Fig. 5.14, where the polarization analogue of the DuMond diagram is shown for a Si crystal and $\Delta\theta_{\mathrm{D}} = \Delta\omega_{\mathrm{D}}/\omega = 10^{-2}$.

Complementary to integral interpretation of the PXR polarization, the fine structure of radiation polarization in the vicinity of the PXR peak has been studied in [65]. The theoretical approach of [65] is based on the kinematic formula (2.29), which connects the PXR spectral–angular distribution of polarization $\boldsymbol{e}_{\mathrm{s}}$, $s = 1, 2$, $(\boldsymbol{e}_{\mathrm{s}} \cdot \boldsymbol{k}) = 0$ proportionally to $(\boldsymbol{e}_{\mathrm{s}}(\omega\boldsymbol{v}/c^2 - \boldsymbol{g}))^2$. For fixed values of the vectors $\boldsymbol{v}, \boldsymbol{g}, \boldsymbol{k}$, the polarization vectors can be chosen to put $\boldsymbol{e}_1$ onto the plane $(k,\, \omega\boldsymbol{v}/c^2 - \boldsymbol{g})$, and $\boldsymbol{e}_2$ perpendicularly to this plane. Then the radiation is locally polarized at each point of the distribution

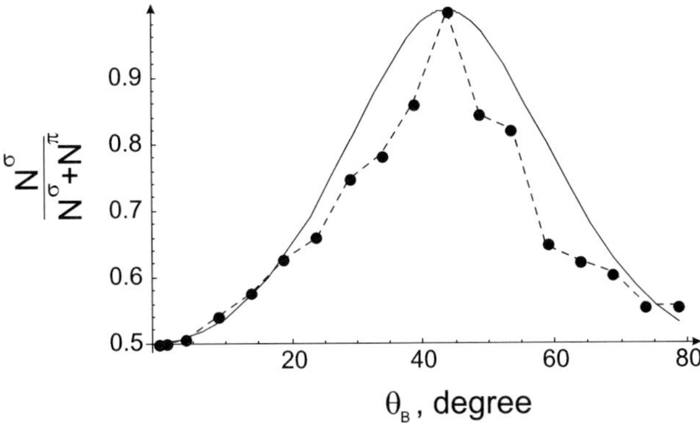


**Fig. 5.14.** Diagrams for the polarization of PXR from the electrons with energy $E = 50\,\text{MeV}$ in a Si crystal for the limiting cases of thin (*solid line*) and thick (*dash line*) crystals

$\theta_x, \theta_y$ within the PXR reflection. The angle between the polarization plane and radiation plane ($\boldsymbol{v}$, $\boldsymbol{k}$) is found in [70]:

$$\Psi = \arctan \frac{\boldsymbol{g}[\boldsymbol{v} \times k]}{(\boldsymbol{vk})(\boldsymbol{g}(\boldsymbol{v}/c + \boldsymbol{k}/k))} \, . \tag{5.38}$$

Figure 5.15 [65] shows the angular distribution of PXR polarization for the (220) reflection in the Si crystal from electrons with energy $E = 80.5\,\text{MeV}$ and

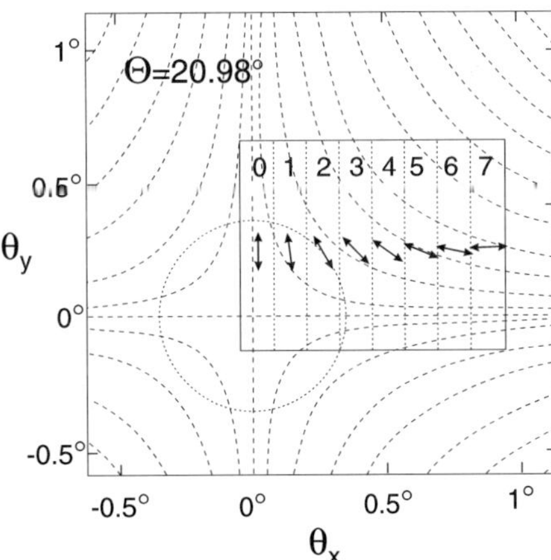


**Fig. 5.15.** Angular distribution of the polarization of PXR [65]. Each *long dash* indicates the direction of the polarization plane calculated using (5.38). Experimental results for the polarization direction and degree, measured within eight stripes denoted by 0–7, are shown by the *bold arrows*. The maximum of the PXR intensity is concentrated in a circle, indicated by *small dashes*

$\theta_B = 20.98°$. The theoretical curves are simulated by (5.38), and the arrows show the experiment [72], where eight independent polarizations have been identified using a CCD polarimeter.

For polarization of radiation from nonrelativistic electrons, the approach [65] can also be applied. However, in this case the wave vector of photons is determined by the detector position (see Chap. 4). Moreover, the interference of PXR and CBS in (4.18) leads to a cumbersome expression for polarization plane parameters, and therefore the formula for this case is skipped here.

## 5.5 Observation of PXR from Protons

The majority of experiments on PXR have been conducted with electron beams due to their wide availability and easy operation. The experiments [3, 14] present the detection of PXR from relativistic protons with energy $E = 70\,\text{GeV}$ at IHEP (Serpukhov, Russia) and $E = 5$ GeV at JINR (Dubna, Russia). The goal of these experiments was to confirm the PXR theory for other than electron particles and to use the PXR phenomenon as an identification procedure for charged particles and evaluation of their energy. Up to now, the paper [14] have been a unique measurement of PXR from heavy particles, and therefore we describe in details the first experiment. In the recent experiment [3], the PXR peaks have been resolved uniquely due to the use of the detector with higher resolution. The monocrystal of silicon $40\times40\times18\,\text{mm}^3$ in size was used for excitation of the PXR (220) reflection. The angle $\theta_B$ was within the range $2°-6°$, and the proton beam of $(1-5) \times 10^6$ particles/fall intensity and $\sim$3 mrad divergence was evacuated from the ring by a bent crystal. The choice of the Bragg angle value was motivated by the optimal crystal thickness for the maximal PXR output in the range of $30-100\,\text{keV}$. The scintillation counter, based on the $\text{Y Al O}_3 : \text{Ce}^{3+}$ crystal of size $25 \times 3$ $\text{mm}^2$, had an relative energy resolution 28% at $\gamma$-line 59.5 keV of collimated source $^{241}\text{Am}$ and had a best efficiency for the mentioned photon energies. The detection scheme registered only those PXR quanta, which were synchronized in time with proton transmission, that improved the signal/noise ratio. The number of protons interacting with target was defined by a plastic scintillation detector placed near the target and included into the synchronization circuit. The distance between the detector and the target was 75 cm.

Figure 5.16 shows the PXR spectra for experiments at two angles $\theta_{B1} = 5.7°$ and $\theta_{B2} = 4.7°$, for which the frequencies (2.34) are $\omega_{B1} \approx 28.2$ keV and $\omega_{B2} \approx 37.6$ keV, respectively. The curves display the obstacle of PXR proton experiments for small observation angles, viz. high background level due to incoherent X-ray radiation from charged particles interacting with the crystal. As the Fig. 5.16 shows ( [14] and private communications with the authors of [14]) the PXR peaks were observed and clearly identified after background substraction.

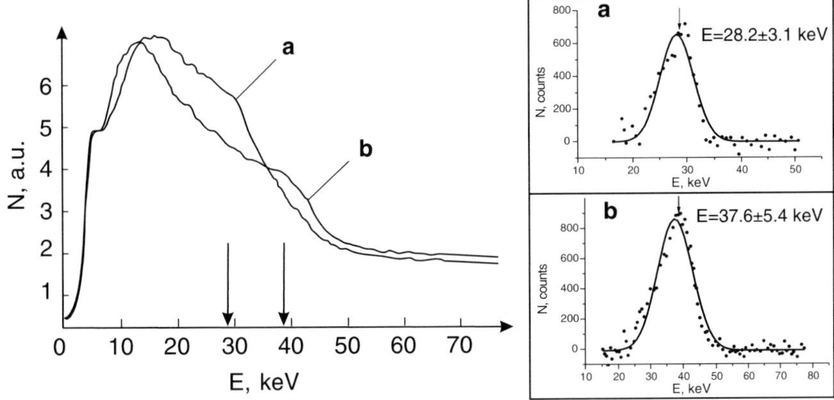


**Fig. 5.16.** PXR spectra from the protons with energy $E = 70$ GeV for the (220) reflection in Si: (**a**) $\theta_{\mathrm{B1}} = 5.7°$; (**b**) $\theta_{\mathrm{B2}} = 4.7°$; the *arrows* show the theoretical values for PXR frequencies $\omega_{\mathrm{B}}$; incuts show the PXR peaks after background substruction

The quantum yield has also been measured for frequencies corresponding to the PXR reflection at the above-mentioned observation angles; $N_{\mathrm{exp}}^{(1)} = (7.87 \pm 1.57) \times 10^{-6}$ quanta/p$^+$ and $N_{\mathrm{exp}}^{(2)} = (1.77 \pm 0.35) \times 10^{-5}$ quanta/p$^+$. For theoretical calculations, formula (2.46) can be used because of essential

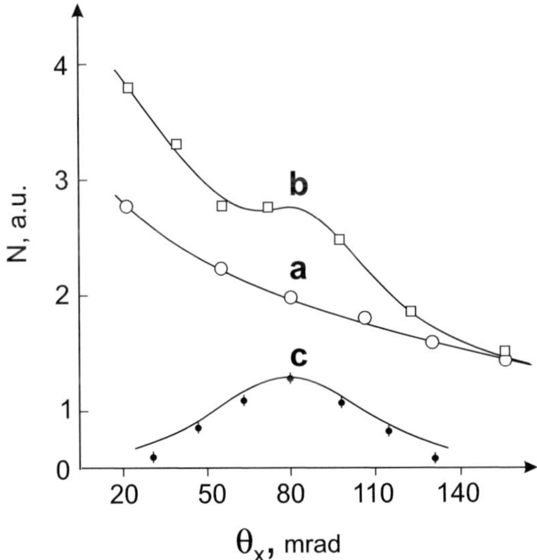


**Fig. 5.17.** Angular distribution of PXR spectra from the protons with energy $E = 70\,\mathrm{GeV}$ for the (220) reflection in Si, $\theta_{\mathrm{B}} = 2.3°$: (**a**) the angle between the proton velocity and plane (220) is $\phi = 0$; (**b**) $\phi = 2.3°$; (**c**) the theoretical distribution; the points correspond to the difference between (a) and (b)

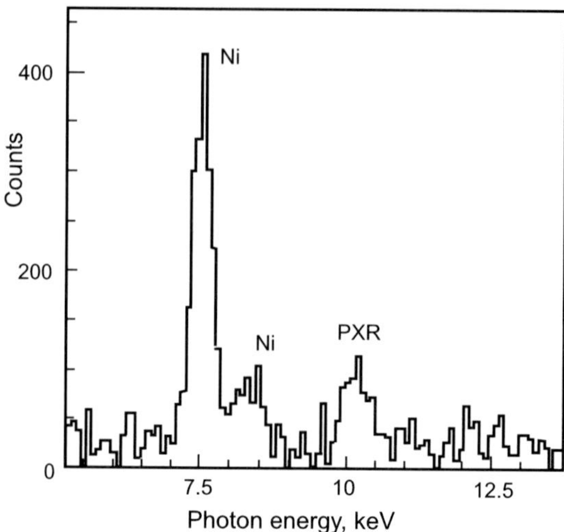


**Fig. 5.18.** Radiation spectrum from the protons of the 5 GeV penetrating silicon crystal. The reflecting plane (100) at $\theta_{\mathrm{B}} = 20°$ has been selected. The clear PXR peak at 10.21 keV along with the characteristic nickel lines is well resolved

suppression of multiple scattering in the case of protons. This calculation [16] results in the following theoretical values: $N^{(1)}_{\mathrm{theor}} = 8.4 \times 10^{-6}$ quanta/p$^+$ and $N^{(2)}_{\mathrm{theor}} = 1.1 \times 10^{-5}$ quanta/p$^+$.

For complete identification of the radiation mechanism, the angular distribution of photons was measured [14] for $\theta_{\mathrm{B}} = 2.3°$, which is shown in Fig. 5.17.

The scanning in the horizontal plane was carried out with the step 25 mrad. Each point in the curve is the result of PXR integration over the energy interval 30–100 keV. The curve in Fig. 5.17a was measured for protons, which incident parallel to the (220) plane, and then the target was rotated to make an angle $\theta_{\mathrm{B}} = 2.3°$ between the planes and beam (the curve in Fig. 5.17b). The results of numerical integration by (2.44) are shown in Fig. 5.17c, where the maximum of radiation is located at $2\theta_{\mathrm{B}} = 4.6° = 80.2$ mrad. According to the theory in Sect. 1.2, the FWHM of PXR angular distribution is conditioned by the energy resolution of the detector, if it is low. This fact explains well the measured value $\Delta\theta_{\mathrm{exp}} = 50 \pm 7$ mrad in the discussed experiment.

The experiment [3] used a semiconductor silicon detector with high resolution, which enhanced the PXR peaks, so they have been observed as distinguishable maxima at low background. The samples were a 100 μm-thick silicon single crystal plate and a 2 mm-thick mosaic pyrolytic graphite specimen. Figure 5.18 shows the radiation spectrum from the (100) Si reflection measured at $\theta_{\mathrm{B}} = 20°$. On the right side from the two characteristic nickel lines (detector housing), the PXR peak at 10.21 keV is clearly seen.

Thus, the experiments [3,14] give a convincing evidence of PXR from heavy particles.

## References

1. D.I. Adeishvili, S.V. Blazhevich, V.F. Boldyshev, G.P. Bochek, V.I. Vitko, V.L. Morochovskii, B.I. Shramenko: Dokl. Akad. Nauk USSR **298**, 844 (1988)
2. Y.N. Adischev, S.N. Arishev, A.V. Vnukov, A.V. Vukolov, A.P. Potylitsyn, S.I. Kuznetsov, V.N. Zabaev, B.N. Kalinin, V.V. Kaplin, S.R. Uglov, A.S. Kubankin, N. Nasonov: Nucl. Instrum. Methods B **201**, 114 (2003)
3. Yu.N. Adishchev, A.S. Artemov, S.V. Afanasiev, V.V. Boiko, M.A. Voevodin, V.I. Volkov, A.S. Gogolev, V.N. Zabaev, A.N. Efimov, Yu.V. Efremov, A.D. Kovalenko, Yu.L. Pivovarov, A.P. Potylitsin, S.V. Romanov, Sh.Z. Saifulin, E.A. Silaev, A.M. Taratin, S.P. Timoshenkov, S.R. Uglov: Russ. JETP Lett. **81**, 305 (2005)
4. Yu.N. Adishchev, R.B. Babadzanov, V.A. Vorob'ev, B.N. Kalinin, V.V. Mun, S. Pak, G.A. Pleshkov, A.P. Potylitsin, S.R. Uglov: Sov. Phys. JETP **66**, 1107 (1987)
5. Y.N. Adishchev, V.G. Baryshevsky, S.A. Vorobiev, V.A. Danilov, S.D. Pak, A.P. Potylizyn, P.F. Safronov, I.D. Feranchuk: Sov. Phys. JETP Lett. **41**, 361 (1985)
6. Yu.N. Adishchev, A.N. Didenko, V.V. Mun, G.A. Pleshkov, A.P. Potylitsin, V.K. Tomchakov, S.R. Uglov, V.A. Vorobiev: Nucl. Instrum. Methods. Phys. Res. B **21**, 49 (1987)
7. Yu.N. Adishchev, V.V. Kaplin, A.P. Potylitsyn, S.R. Uglov, S.A. Vorobyev, R.D. Babadzanov, V.A. Verzilov: Phys. Lett. A **147**, 326 (1990)
8. Yu.N. Adishchev, V.A. Verzilov, A.P. Potylitsyn, S.R. Uglov, S.A. Vorobyev: Nucl. Instrum. Methods B **44**, 130 (1989)
9. Yu.N. Adishchev, V.A. Verzilov, S.A. Vorobyev, A.P. Potylitsyn, S.R. Uglov: JETP Lett. **48**, 342 (1988)
10. Y.N. Adischev, V.N. Zabaev, V.V. Kaplin, S.V. Razin, S.R. Uglov, S.I. Kuznetzov, Yu.P. Kunashenko: Nucl. Phys. (rus) **66**, 446 (2003)
11. V.P. Afanasenko, V.G. Baryshevsky, S.V. Gatsicha, R.F. Zuevsky, M.G. Lifshits, A.S. Lobko, V.I. Moroz, V.V. Panov, I.V. Polikarpov, V.P. Potsiluiko, P.F. Safronov, A.O. Yurtsev: Sov. JETP Lett. **51**, 213 (1990)
12. V.P. Afanasenko, V.G. Baryshevsky, O.T. Gradovsky, R.F. Zuevsky, M.G. Livshits, A.S. Lobko, V.I. Moroz, V.V. Panov, I.V. Polikarpov, D.S. Shvarkov, A.O. Yurtsev: Phys. Lett. A **141**, 311 (1989)
13. V.P. Afanasenko, V.G. Baryshevsky, A.S. Lobko, V.V. Panov, R.F. Zuevsky: Nucl. Instrum. Methods A **334**, 631 (1993)
14. V.P. Afanasenko, V.G. Baryshevsky, R.F. Zuevsky, A.S. Lobko, A.A. Moskatelnikov, V.V. Panov, V.P. Potsiluiko, S.V. Skorochod, D.S. Shvarkov: Phys. Lett. A **170**, 315 (1992); Sov. JETP Lett. **54**, 493 (1991)
15. A.I. Akhiezer, V.B. Berestetzkii: *Quantum Electrodynamics* (Nauka, Moscow 1969) pp 463–467
16. T. Akimoto, M. Tamura, J. Ikeda, Y. Aoki, F. Fujita, K. Sato, A. Honma, T. Sawamura, M. Narita, K. Imai: Nucl. Instrum. Methods A **459**, 78 (2001)
17. A.N. Aleinik, A. Saryshev, E.A. Bogomazova, B.N. Kalinin, G.A. Naumenko, A.P. Potylitsin, G.A. Saruev, A.F. Sharafutdinov: JETP Lett. **79**, 396 (2004)
18. K.Yu. Amosov, B.N. Kalinin, A.P. Potylitsin, V.P. Sarychev, S.R. Uglov, V.A. Verzilov, S.A. Vorobiev, I. Endo, T. Kobayashi: Phys. Rev. E **47**, 2207 (1993)
19. A.V. Andreev: Usp. Fiz. Nauk **145**, 113 (1985)

20. M.Yu. Andreyashkin, V.V. Kaplin, A.P. Potylitsin, S.R. Uglov, V.N. Zabaev, K. Aramitsu, I. Endo, K. Goto, T. Horiguchi, T. Kobayashi, Y. Takashima, M. Muto, K. Yoshida, H. Nitta: Nucl. Instrum. Methods B **119**, 108 (1996)
21. M.Yu. Andreyashkin, V.V. Kaplin, S.R. Uglov, V.N. Zabaev, M. Piestrup: Appl. Phys. Lett. **72**, 1385 (1998)
22. M.Yu. Andreyashkin, V.N. Zabaev, V.V. Kaplin, K. Nakayama S.R. Uglov, I. Endo: JETP Lett. **65**, 594 (1997)
23. M.Yu. Andreyashkin, V.N. Zabaev, K. Yoshida, V.V. Kaplin, E.I. Rozum, S.R. Uglov, I. Endo: JETP Lett. **42**, 770 (1995)
24. S. Asano, I. Endo, M. Harada, S. Ishii, T. Kobayashi, T. Nagata, M. Muto, K. Yoshida, H. Nitta: Phys. Rev. Lett. **70**, 3247 (1993)
25. A. Authier: *Dynamical Theory of X-ray Diffraction* (Oxford University Press, New York 2001) pp 3–26
26. R.O. Avakian, A.E. Avetisian, Yu.N. Adishchev, G.M. Garybian, S.S. Danagulian, O.S. Kizogian, A.P. Potylitsin, S.P. Taroian, G.M. Elbakian, Yan Shi: JETP Lett. **45**, 313 (1987)
27. H. Backe, C. Ay, N. Clawiter, Th. Doerk, M. El-Ghazaly, K.-H. Kaiser, O. Kettig, G. Kube, F. Hagenbuck, W. Lauth, A. Rueda, A. Sharafutdinov, D. Schroff, T. Weber: *Proc. Int. Symp. on Channelling – Bent Crystals – Radiation Processes* (J.W. Goethe University, Frankfurt am Main 2003), p 41
28. H. Backe, A. Rueda, W. Lauth, N. Clawiter, M. El-Ghazaly, R. Kunz, T. Weber: Nucl. Instrum. Methods B, **234**, 138 (2005)
29. V.N. Baier, V.M. Katkov, V.M. Strakhovenko: *Electromagnetic Processes at High Energies in Oriented Single Crystals* (World Scientific, Singapore 1998)
30. V.G. Baryshevsky, V.A. Danilov, O.L. Ermakovich, I.D. Feranchuk, A.V. Ivashin, V.I. Kozus, S.G. Vinogradov: Phys. Lett A **110**, 477 (1985)
31. V.G. Baryshevsky, I.D. Feranchuk: Dokl. Belarussian Akad. Nauk **27**, 995 (1983)
32. V.G. Baryshevsky, I.D. Feranchuk: Dokl. Belarussian Akad. Nauk **28**, 336 (1984)
33. V.G. Baryshevsky, I.D. Feranchuk: Phys. Lett. **A 102**, 141 (1984)
34. V.G. Baryshevsky, I.D. Feranchuk: Izv. Belarussian Akad. Nauk, Ser. Fiz-Mat. Nauk **2**, 79 (1985)
35. V.G. Baryshevsky, I.D. Feranchuk, A.O. Grubich, A.V. Ivashin: Nucl. Instrum. Methods A **249**, 306 (1986)
36. V.G. Baryshevsky, A.O. Grubich, I.D. Feranchuk: Zh. Eksp. Teor. Fiz. **90**, 1588 (1986)
37. S.V. Blazhevich, G.L. Bochek, V.B. Gavrikov, V.I. Kulibaba, N.I. Maslov, N.N. Nasonov, B.H. Pyrogov, A.G. Safronov and A.V. Torgovkin: Phys. Lett. A **195**, 210 (1994)
38. M. Born, E. Wolf: *Principles of Optics* (Pergamon Elmsford, New York 1975) p 328
39. K.-H. Brenzinger, C. Herberg, B. Limburg, H. Backe, S. Dambach, H. Euteneuer, F. Hagenbuck, H. Hartmann, K. Johann, K.H. Kaiser, O. Kettig, G. Knies, G. Kube, W. Lauth, H. Schöpe, Th. Walcher: Zs. f. Phys. A **358**, 107 (1997)
40. K.-H. Brenzinger, B. Limburg, H. Backe, S. Dambach, H. Euteneuer, F. Hagenbuck, C. Herberg, K.H. Kaiser, O. Kettig, G. Kube, W. Lauth, H. Schöpe, Th. Walcher: Phys. Rev. Lett. **79**, 2462 (1997)
41. A. Caticha: Phys. Rev. A **40**, 4322 (1989); Phys. Rev. B **45**, 9541 (1992)
42. A. Caticha, S. Caticha-Ellis: Phys. Rev. B **25**, 971 (1982)
43. A.N. Didenko, Yu.N. Adishchev, B.N. Kalinin, A.P. Potylitsin, S.A. Vorobiev, V.V. Mun and S. Pak: Phys. Lett. A **118**, 363 (1986)

44. A.N. Didenko, B.N. Kalinin, S. Pak, A.P. Potylitsin, S.A. Vorobiev, V.G. Baryshevsky, V.A. Danilov, I.D. Feranchuk: Phys. Lett. A **110**, 177 (1985)
45. I. Endo, M. Harada, T. Kobayashi, Y.S. Lee, T. Ohgaki, T. Takahashi, M. Muto, K. Yoshida, H. Nitta, T. Ohba: Phys. Rev. E **51**, 6305 (1995)
46. I.D. Feranchuk: Coherent phenomena in the processes of X-ray and gamma-radiation from relativistic charged particles in crystals. Habilitation Doctor Thesis, Belarussian State University, Minsk (1984)
47. I.D. Feranchuk, L.I. Gurskii, L.I. Komarov, O.M. Lugovskaya, F. Burgäzy, A.P. Ulyanenkov: Acta Crystallogr. A **58**, 370 (2002)
48. I.D. Feranchuk, A.V. Ivashin: J. Phys. (Paris) **46**, 1981 (1985)
49. R.B. Fiorito, D.W. Rule, X.K. Maruyama, K.L. DiNova, S.J. Evertson, M.J. Osborne, D. Snyder, H. Rietdyk, M.A. Piestrup, A.H. Ho: Phys. Rev. Lett. **71**, 704 (1993)
50. R.B. Fiorito, D.W. Rule, M.A. Piestrup, Q. Li, A.H. Ho, X.K. Maruyama: Nucl. Instrum. Methods B **79**, 758 (1993)
51. R.B. Fiorito, D.W. Rule, M.A. Piestrup, X.K. Maruyama, R.M. Silzer, D.M. Skopik, A.V. Shchagin: Phys. Rev. E **51**, R2759 (1995)
52. J. Freundenberger, M. Galemann, H. Genz, L. Groening, P. Hoffmann-Stascheck, V.L. Morokhovskii, V.V. Morokhovskii, U. Nething, H. Prade, A. Richter, J.P.F. Sellschop, R. Zahn: Nucl. Instrum. Methods **115**, 408 (1996)
53. J. Freundenberger, V.B. Gavrikov, M. Galemann, H. Genz, L. Groening, V.L. Morokhovskii, V.V. Morokhovskii, U. Nething, A. Richter, J.P.F. Sellschop, N.F. Shul'ga: Phys. Rev. Lett. **74**, 2487 (1995)
54. J. Freundenberger, H. Genz, V.V. Morokhovskii, A. Richter, J.P.F. Sellschop: Phys. Rev. Lett. **84**, 270 (2000)
55. Y. Hayakawa, M. Seto, Y. Maeda, T. Shirai, A. Noda: J. Phys. Soc. Japan **67**, 1044 (1998)
56. R.W. James: *The Optical Principles of the Diffraction of X-rays* (G. Bell and Sons, London 1950)
57. V.M. Kaganer, V.L. Indenbom, M. Vraha, B. Chalupa: Phys. Status Solidi a **71**, 372 (1982)
58. B.N. Kalinin, G.A. Naumenko, D.V. Padalko, A.P. Potylitsyn, I.E. Vnukov: Nucl. Instrum. Methods B **173**, 253 (2001)
59. V.V. Kaplin, S.I. Kuznetsov, N.A. Timchenko, S.R. Uglov, V.N. Zabaev: Nucl. Instrum. Methods B **173**, 238 (2001)
60. V.V. Kaplin, S.R. Uglov, V.N. Zabaev, V.G. Kanaev, S.V. Litvin and N.A. Timchenko: Nucl. Instrum. Methods A **448**, 66 (2000)
61. V.V. Kaplin, S.R. Uglov, V.N. Zabaev, M.A. Piestrup, C.K. Gary, N.N. Nasonov, M.K. Fuller: Appl. Phys. Lett. **76**, 3647 (2000)
62. Y.S. Korobochko, V.F. Kosmach, V.I. Mineev: Sov. Phys. JETP **21**, 834 (1965)
63. O.M. Lugovskaya, I.Ya. Dubovskaya: Yad. Fiz., **66**, 429 (2003) (Phys. At. Nuclei **66**, 404 (2003))
64. V.V. Morokhovskyi, J. Freundenberger, H. Genz, V.L. Morokhovskii, A. Richter, J.P.F. Sellschop: Phys. Rev. B **61**, 3347 (2000)
65. V.V. Morokhovskii, J. Freudenberger, H. Genz, A. Richter, K.H. Schmidt, G. Buschhorn, R. Kotthaus, M. Rzepka, P.M. Weinmann: Nucl. Instrum. Methods B **145**, 14 (1998)
66. V.V. Morokhovskii, K.H. Schmidt, G. Buschhorn, J. Freundenberger, H. Genz, R. Kotthaus, A. Richter, M. Rzepka, P.M. Weinmann: Phys. Rev. Lett. **79**, 4389 (1997)

67. V.V. Morokhovskii, K.H. Schmidt, G. Buschhorn, J. Freundenberger, H. Genz, R. Kotthaus, A. Richter, M. Rzepka, P.M. Weinmann: Phys. Rev. Lett. **79**, 4389 (1997)
68. M.A. Piestrup, X. Wu, V.V. Kaplin, S.R. Uglov, J.T. Cremer, D.W. Rule, R.B. Fiorito: Rev. Sci. Instrum. **72**, 2159 (2001)
69. M.A. Porai-Koshitz: *Practical Course for X-ray Structure Analysis, vol 2* (Fizmatgiz, Moscow 1968)
70. D. Pugachov, G. Buschhorn, R. Kotthaus, V.L. Morokhovskii, J. They, H. Genz, A. Richter, A. Ushakov: Phys. Lett. A **286**, 70 (2001)
71. G.M. Reese, J.C.H. Spence, N. Yamamoto: Phil. Mag. A **49**, 697 (1984)
72. K.H. Schmidt, G. Buschhorn, R. Kotthaus, M. Rzepka, P.M. Weinmann, V.V. Morokhovskii, J. Freudenberger, H. Genz, A. Richter: Nucl. Instrum. Methods B **145**, 8 (1998)
73. A.V. Shchagin, V.I. Pristupa, N.A. Khizhnyak: Phys. Lett. A **148**, 485 (1990)
74. B. Sones, Y. Danon, R.C. Block: Nucl. Instrum. Methods B **227**, 22 (2005)
75. T. Takahashi, Y. Shibata, K. Ishi, M. Ikezawa, M. Oyamada, Y. Kondo: Phys. Rev. E **62**, 8606 (2000)
76. M.L. Ter-Mikaelian: *High Energy Electromagnetic Processes in Condensed Media* (in Russian: AN ArmSSR, Yerevan 1969) (in English: Wiley, New York 1972)
77. A. Ulyanenkov, I.D. Feranchuk, L.I. Komarov: 50th DXC Proceedings, Steamboat Springs, p 55 (2001)
78. A. Vorob'ev, B.N. Kalinin, S. Pak, A.P. Potylitsin: Sov. Phys. JETP Lett. **41**, 1 (1985)
79. I.E. Vnukov, B.N. Kalinin, G.A. Naumenko, D.V. Padalko, A.P. Potylitzin: Fiz., Izv. VUZov **44**, 53 (2001)

# 6

# High-Resolution PXR Experiments

## 6.1 Spectral Width of PXR Peaks

The PXR experiments discussed in the previous chapter are referred to as low-resolution scale (LRS) measurements, according to the classification in Sect. 3.4. The results reported in various publications on LRS demonstrate the applicability of the kinematic diffraction theory and simple averaging model for multiple scattering and mosaicity for the treatment of experimental data.

Recently, several high-resolution scale PXR experiments have been conducted, which opens new application abilities for this phenomenon. The theoretical treatment of HRPXR is based on the dynamical diffraction theory (see Chap. 3). The first measurements of the PXR natural linewidth were performed in [20] applying an absorption technique. However, the low-energy electrons with $E = 10$ MeV were used in this experiment, and the PXR linewidth was found as 48 eV for the photons of $\hbar\omega = 8.98$ keV, which refers in this case to LRS. The new qualitative level was reached by using a double-crystal set-up, widely used in conventional X-ray diffraction and utilized for PXR for the first time in [7]. This experiment was conducted at Mainz Microtron MAMI with 855 MeV electrons. The experimental set-up is depicted in Fig. 6.1. The electron beam of divergence $\sim 35$ μrad was evacuated from the storage ring by the magnetic field and a struck Si single crystal of thickness $L = 525$ μm. The same crystal with the thickness $L_1 = 1$ mm was used as an analyser. The geometry of the experiment was the Bragg symmetric reflection from the family of $\{111\}$ planes; the variation of $\theta_B = (\pi - \psi_{1,x})/2$ was achieved by rotating a crystal by ($-10$ to $+10$) mrad. The reflecting planes of the analyser crystal were set to the energy dispersive position [22] with the shift $\Delta\psi_{2,x} = 45.6$ mrad, as shown in the DuMond diagram (Fig. 6.1b). This geometry allowed us to scan an energy spectrum in the range ($-0.3$–$0.3$) eV with respect to the centre of the reflection $\hbar\omega_B^{(l)}$.

Four reflections were investigated in [7], for which the frequency is calculated from (3.17):

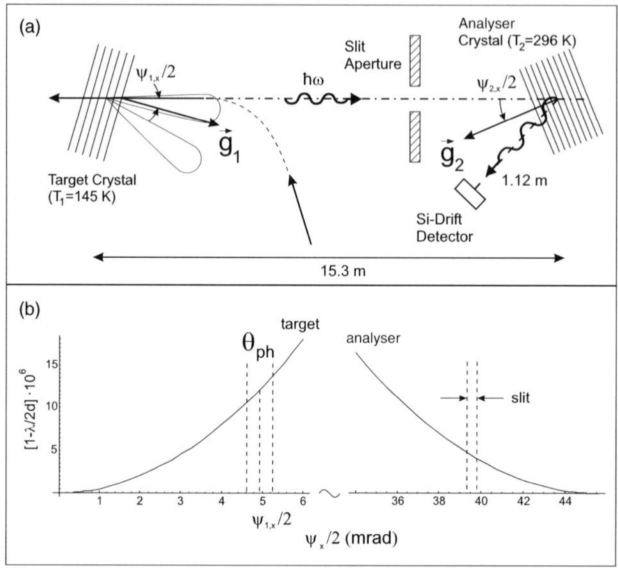


**Fig. 6.1.** Dispersive double-crystal arrangement for observation of radiation in the backward geometry: (**a**) experimental set-up from [7], rotation of the target crystal permitted to change the angle $\theta_{\rm B} = (\pi - \psi_{1,x})/2$, reflection from the analyser crystal permitted to scan the photon energy spectrum by changing the tilt angle $\psi_{2,x}/2$; (**b**) corresponding to (a) DuMond diagram

$$\hbar\omega_{\rm B}^{(l)} = \hbar\omega_{(111)} l, \qquad l = 1, 3, 4, 5 \; ;$$
$$\hbar\omega_{(111)} = 1.974 \frac{\pi\sqrt{3}}{a_{\rm Si}} = 1.978 \text{ keV} \; . \tag{6.1}$$

Table 6.1 contains the parameters required for theoretical interpretation of data from [7]. The thickness of the target crystal satisfies

$$L > L_{\rm abs} \gg L_{\rm ext} \; ,$$

i.e. the formulas from Sect. 3.4 can be used, and the radiation intensity does not depend on $L$. Moreover, the multiple scattering can be neglected in simulations, because a Bragg diffracted wave is formed [6] on the limited length of the crystal ($\sim L_{\rm ext}$), and for the electron energy $E = 855\,$MeV condition (1.26) is fulfilled:

$$\theta_{\rm s}^2(L_{\rm ext}) = \frac{E_{\rm s}^2}{E^2}\frac{L_{\rm ext}}{L_R} < \gamma^{-2} \; .$$

For the angle $\theta_{\rm B} \approx \pi/2$, the parameters $|C_{\rm s}| = 1, \beta = -1$, and the spectral and angular variables for photons arrived at the analyser, are related to the rotation angle of the target crystal as

$$\theta_x \approx \psi_{1,x}, \qquad \theta_y \approx 0, \qquad \alpha_{\rm B} \approx -4\frac{\omega - \omega_{\rm B}^{(l)}}{\omega_{\rm B}^{(l)}} + 2\psi_{1,x}^2 \; , \tag{6.2}$$

**Table 6.1.** Parameters for data interpretation [7]

| $(hhh)$ | $\gamma^{-2}$ | $\theta_s^2(L)$ | $\lvert\chi_0'\rvert \times 10^5$ | $\chi_0'' \times 10^6$ | $\lvert\chi_L\rvert \times 10^5$ | $\hbar\omega_B$ (keV) |
|---|---|---|---|---|---|---|
| (111) | | | 20.8 | 69.1 | 11.3 | 1.97 |
| (333) | $3.6 \times 10^{-7}$ | $3.1 \times 10^{-6}$ | 2.89 | 1.20 | 0.97 | 5.93 |
| (444) | | | 1.60 | 0.39 | 0.59 | 7.91 |
| (555) | | | 1.03 | 0.16 | 0.20 | 9.89 |

with the accuracy of detector slit parameters $\delta_x, \delta_y$. The radiation is polarized within the plane $(\boldsymbol{v}, \boldsymbol{g})$ and (3.29)–(3.31) deliver the spectral–angular distribution of single PXR reflection:

$$\frac{\partial^2 N_\pi^{(l)}}{\partial \psi_{1,x} \partial \omega} = \frac{\alpha \delta_y}{\pi^2 \omega_B^{(l)}} \frac{(\psi_{1,x}^2 + 1/2\delta_y^2) \left| \chi_g(\omega_B^{(l)}) \right|^2}{|\epsilon(\alpha_B)|^2}$$

$$+ \left| \frac{1}{\psi_{1,x}^2 + \gamma^{-2}} - \frac{1}{\psi_{1,x}^2 + \gamma^{-2} - \chi_0(\omega_B^{(l)}) - \epsilon(\alpha_B)} \right|^2 ,$$

$$\epsilon(\alpha_B) = \frac{1}{2} \left[ \mu + \text{sign}\, \mu \sqrt{\mu^2 - 4\chi_g \chi_{-g}} \right] , \qquad \mu = \alpha_B - 2\chi_0 \, . \tag{6.3}$$

The accuracy of (6.3) is adjusted to $\delta_y \ll \gamma^{-1}$. The parameter $\mu/2|\chi_g|$ is equivalent to the deviation parameter for the reflection curve in the dynamical theory [6]. Using (6.3), we discuss here a physical meaning of amplitudes of the processes forming the total PXR intensity (see also Sect. 3.4). For this purpose, the variable $\mu_r$ for the maximum position and natural angular $\Delta\Psi$ and spectral $\Gamma$ widths are introduced to characterize the amplitudes by taking part in the formation of PXR reflection [13, 17].

In the first term of (6.3), the angular and spectral distributions are not interdependent and, therefore, a simple estimates for them are

$$\mu_r^{(1)} \approx 0, \quad \Delta\Psi_1^{(l)} \approx \gamma^{-1}, \quad \Gamma_1^{(l)} \approx 2|\chi_g|\omega_B^{(l)}, \quad N_{1m}^{(l)} \approx \frac{\alpha\delta_y}{\pi^2} \frac{\psi_{1,x}^2}{\psi_{1,x}^2 + \gamma^{-2}}, \tag{6.4}$$

where $N_{1m}^{(l)}$ determines the intensity of the spectral maximum of the amplitude. The above parameters describe the radiation which is formed on the extinction length $L_{\rm ext}$ and can be considered as a Bragg diffraction of the electromagnetic field of a particle, or diffraction transition radiation (DTR) [7].

The position of the spectral maximum of the amplitude in the second term depends on both angle and reflection order:

$$\psi_{1,x}^2 + \gamma^{-2} = \text{Re} \left\{ \chi_0(\omega_B^{(l)}) - \epsilon(\alpha_B) \right\} \, . \tag{6.5}$$

Here, the radiation is formed on the absorption length $L_{\rm abs}$ and corresponds to the Cherenkov mechanism (VCR). The parameters in this case are

$$\mu_{\rm r}^{(2)} \approx \psi_{1,x}^2 + \gamma^{-2} + \left|\chi_0(\omega_{\rm B}^{(l)})\right| \,,$$
$$\Delta\Psi_2^{(l)} \approx \sqrt{|\chi_0(\omega_{\rm B}^{(l)})|} \qquad \Gamma_2^{(l)} \approx {\rm Im}\{\chi_0\}\omega_{\rm B}^{(l)} \ll \Gamma_1^{(l)},$$
$$N_{2m}^{(l)} \approx \frac{\alpha\delta_y}{\pi^2} \frac{\psi_{1,x}^2}{\psi_{1,x}^2 + \gamma^{-2} + \left|\chi_0(\omega_{\rm B}^{(l)})\right|} \left|\frac{\chi_g(\omega_{\rm B}^{(l)})}{{\rm Im}\{\chi_0(\omega_{\rm B}^{(l)})\}}\right|^2 . \tag{6.6}$$

Equations (6.2)–(6.6) prove the conditional character of mechanism separation in the formation of PXR, because the amplitude ratio and interference have a dynamical nature and depend on the reflection. This fact is illustrated by theoretical interpretation of the experiment [7], which was carried out with the one-crystal scheme (LRS mode). The energy resolution of the semiconductor or CCD detector was $\hbar\Delta\omega_{\rm D} \approx (0.2–0.4)$ keV, which is enough to distinguish between the different reflections $l$. Due to validity of the condition $\Delta\omega_{\rm D} \gg \Gamma_{1,2}^{(l)}$, the integration over the frequency in (6.3) can be performed:

$$\frac{\partial N_\pi^{(l)}}{\partial\psi_{1,x}} = \frac{2\alpha\delta_y\left|\chi_g(\omega_{\rm B}^{(l)})\right|}{\pi^2} \frac{(\psi_{1,x}^2 + 1/2\delta_y^2)}{(\psi_{1,x}^2 + \gamma^{-2})^2} I(\psi_{1,x}) \,,$$
$$I(\psi_{1,x}^2) = \int_{-\infty}^{\infty} \frac{{\rm d}t}{|t + {\rm sign}\, t\sqrt{t^2-1}|^2}$$
$$+ \left|1 - \frac{\psi_{1,x}^2 + \gamma^{-2}}{\psi_{1,x}^2 + \gamma^{-2} - \chi_0 - \chi_g(t + {\rm sign}\, t\sqrt{t^2-1})}\right|^2 . \tag{6.7}$$

The integral (6.7) is evaluated analytically, yet results in a cumbersome expression. Figure 6.2 shows the simulated PXR angular distribution using (6.7). The radiation intensity is non-zero even for $\psi_{1,x} = 0$ due to finite angular resolution of slit $\delta_y$.

The high-resolution part of the experiment [7] was carried out with the double-crystal scheme (Fig. 6.1). The radiation frequency spectrum was measured at two angles $\psi_{1,x}^{(1)}/2 = 0.3$ mrad and $\psi_{1,x}^{(2)}/2 = 5$ mrad. The theoretical shape of the natural PXR lines is calculated for different orders of $\{111\}$ series in [8,13] on the basis of the dynamical theory and neglecting the multiple scattering ((6.3) along with the data in Table 6.1 can be used), and the results for [7] are shown in Fig. 6.3.

The difference in the position and the width of HRPXR lines in the left and right panels of Fig. 6.3 is explained by (6.4)–(6.6), if the following inequality is used:

$$\psi_{1,x}^{(1)}/2 \approx \gamma^{-1} \ll \psi_{1,x}^{(2)}/2 \approx \sqrt{|\chi_0(\omega_{\rm B}^{(l)})|} \,.$$

For small values of $\psi_{1,x}^{(1)}/2$, the maxima of DTR and VCR amplitudes are of the same order. On the other hand, for $\psi_{1,x}^{(2)}/2$ the shift of the VCR line with respect to the reflection centre depends on the index $l$, and the amplitude of DTR decreases by $\left(\gamma\psi_{1,x}^{(2)}/2\right)^{-2} \sim 10^{-2}$ times.

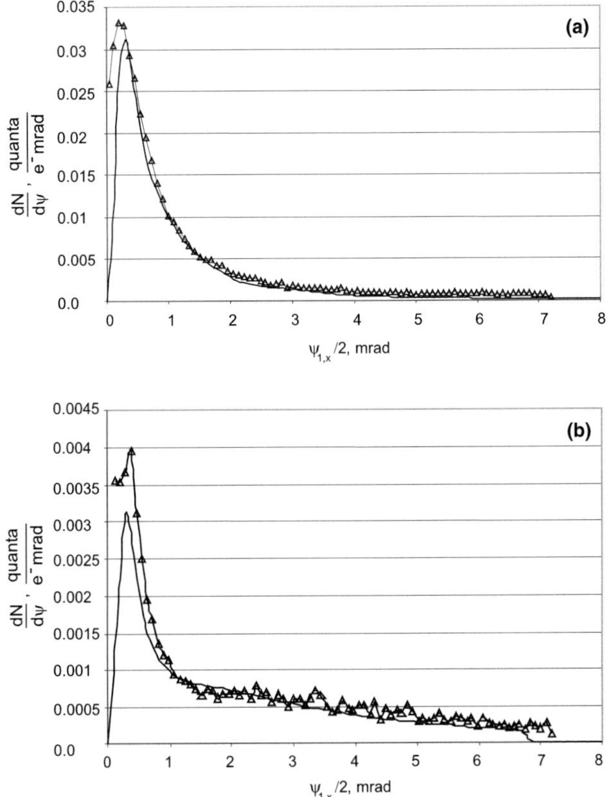


**Fig. 6.2.** PXR angular distribution for the experimental parameters from [7]: (**a**) the (111) reflection; (**b**) the (333) reflection. The *dashed lines* correspond to the experimental data [7], the *solid lines* represent the calculation by (6.7) and the *dotted lines* shows the contribution of DTR alone

Figure 6.4 shows the HRPXR measurements from [7]. The position of peaks fits well to theoretical predictions; however, the dynamics of their width is essentially different from the expected behaviour of natural lines. This effect also takes place in HRXRD [6], where the line shape is the result of convolution of natural spectrum with the instrumental function of the slit and the analyser. Since in [7] the same Si crystal for the analyser was used as as for the target, the deviation parameter $\mu/2|\chi_g|$ is equivalent for both. Therefore, using the simple approximation for the instrumental function, the line shape is found to be Gaussian with the FWHM depending on resolutions of all instrumental elements:

$$\Delta\left(\hbar\omega_{1,2}^{(l)}\right) \approx \hbar\sqrt{\left[\Gamma_{1,2}^{(l)}\right]^2 + \left[2|\chi_g|\omega_{\mathrm{B}}^{(l)}\right]^2 + \left[\omega_{\mathrm{B}}^{(l)}\right]^2 \theta_s^2}\,. \tag{6.8}$$

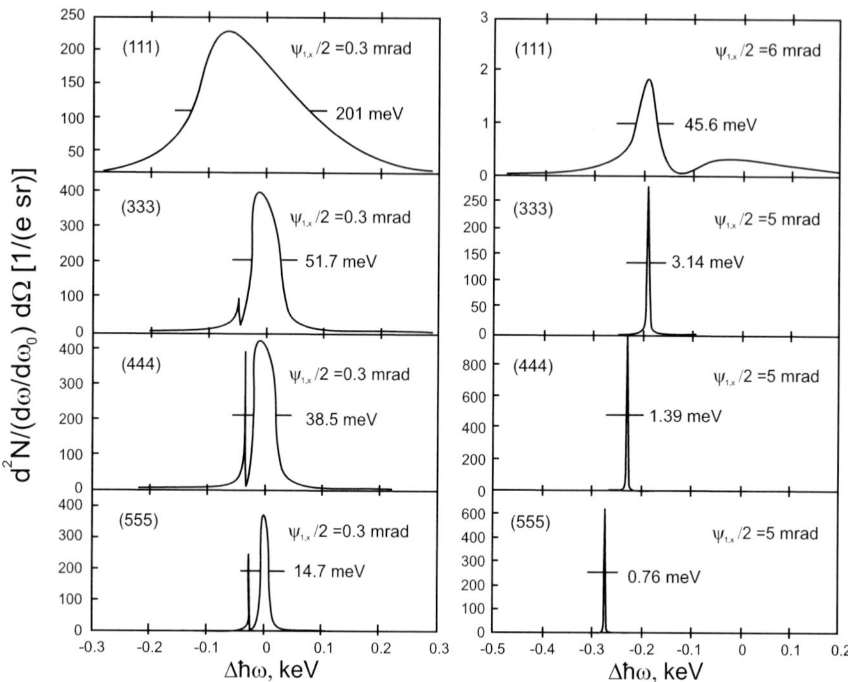


**Fig. 6.3.** Calculated spectral distributions of X-rays emitted from a semi-infinite Si single crystal in the backward direction for an electron with an energy of 855 MeV [7]. The left panel shows the results for various reflection orders at the angle $\psi_{1,x}^{(1)}/2$, and the right panel shows the results at the angle $\psi_{1,x}^{(2)}/2$

Here, $\Gamma_{1,2}^{(l)}$ are the spectral widths of PXR, the second term in (6.8) results from the Darwin–Prins reflection of the analyser, and the third term determines the mean-square angle of multiple scattering, $\theta_s^2$.

Figure 6.5 shows an agreement between the measured [7] and calculated values of FWHM by (6.8) ($\theta_s \approx 1.7$ mrad) at different $\psi_{1,x}$.

## 6.2 Forward Direction PXR

The specific character of the photon spectrum in the direction of movement of relativistic charged particles was pointed out in the first theoretical works on PXR [10, 12]. This specificity is caused by the dynamical character of the formation of radiation in the crystal, and this fact is well explained by the classic (2.16) or quantum (3.8) theory. In contrast to a homogeneous medium, where the X-ray intensity is determined by a plane wave, in a crystal the radiation is formed by the vector potential $\boldsymbol{A}_{\boldsymbol{k}s}(\boldsymbol{r})$ for Bloch states of the electromagnetic field. The Bloch states are the stationary superpositions of

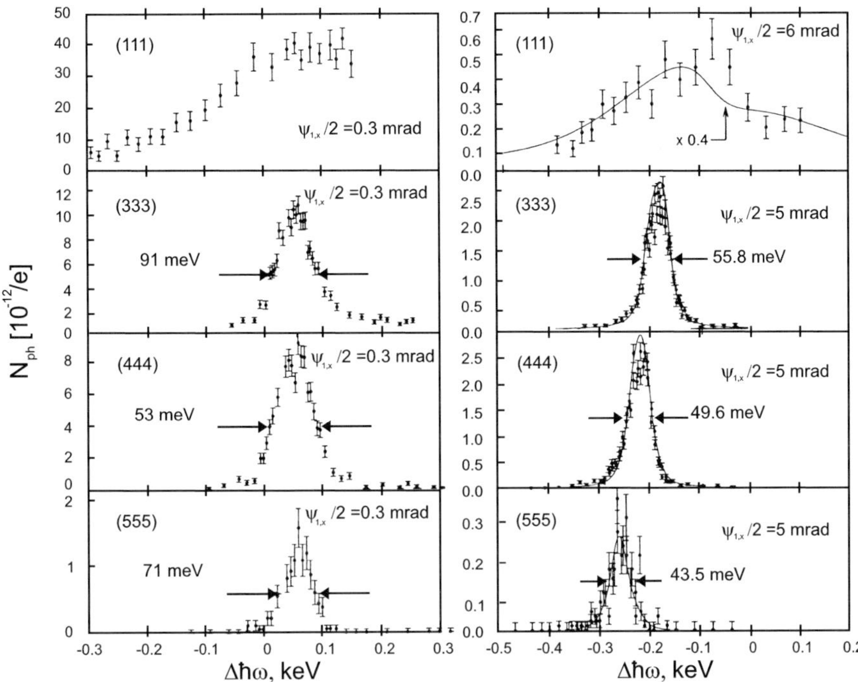


**Fig. 6.4.** Experimental data for line shape measurements with double-crystal arrangement from [7]. Left and right panels correspond to the same cases an in Fig. 6.3

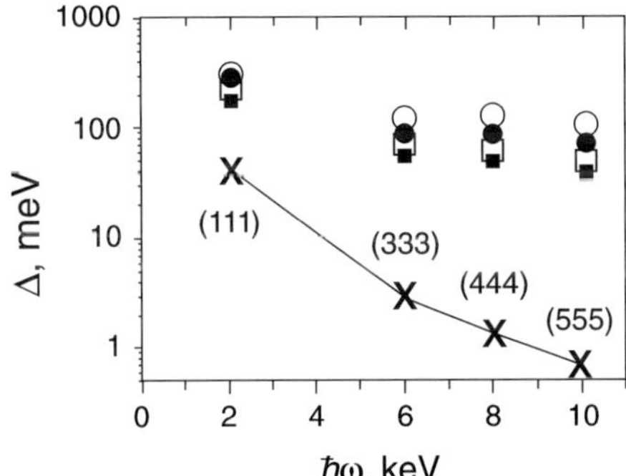


**Fig. 6.5.** Measured and calculated values for the PXR linewidths: the black circles (experiment) and open circles (theory) correspond to the electron tilt angle $\psi_{1,x}/2 =$ 0.3 mrad; the black squares and open squares show the values in the case of $\psi_{1,x}/2 =$ 5.0 mrad; $\times$ corresponds to the natural bandwidth of the PXR line $\Gamma_2^{(l)}$

direct and diffracted waves, which in the two-beam approximation are ($\mu = 1, 2$)

$$\boldsymbol{A}_{\boldsymbol{k}s}^{(\mu)}(\boldsymbol{r}) = \mathrm{e}^{\mathrm{i}\boldsymbol{k}\boldsymbol{r}n_{\mu s}} \left[ \boldsymbol{e}_{\mathrm{s}} A_0^{(\mu)} + \boldsymbol{e}_{gs} A_g^{(\mu)} \mathrm{e}^{\mathrm{i}\boldsymbol{g}\boldsymbol{r}} \right] . \tag{6.9}$$

The effective refraction index $n_{\mu s} = 1+2\epsilon_{\mu s}$ for the electromagnetic wave in a crystal, and the polarization vectors $\boldsymbol{e}_{\mathrm{s}}, \boldsymbol{e}_{1s}$ and the relation between $A_0^{(\mu)}, A_g^{(\mu)}$ are defined in (3.14)–(3.16). Expression (6.9) is valid for any $\boldsymbol{k}$ [6]; however, the amplitude $A_g^{(\mu)}$ becomes essential near the Bragg condition (3.26):

$$|\alpha_{\mathrm{B}}| = \left| \frac{2\boldsymbol{k}\boldsymbol{g} + g^2}{k^2} \right| \leq |\chi_g| . \tag{6.10}$$

Because the direct and diffracted waves in the crystal are independent, the change of the amplitude of one wave near (6.10) causes the change of another. In the dynamical diffraction theory, this connection leads to the known pendellösung effect [6]. By analogy, the emission of PXR from a relativistic particle into side reflection (amplitude $A_g^{(\mu)}$) influences the spectrum (amplitude $A_0^{(\mu)}$) of the forward direction PXR (FDPXR). Contrary to PXR at large angles to the particle velocity, this phenomenon cannot be described by the first-order perturbation theory on crystal polarizability [26].

The mechanism of FDPXR differs from resonant transition or diffracted radiation [26]. The latter is an interference of transition radiation from crystallographic planes; its optical equivalent is the Smith–Purcell radiation [24]. This radiation (see Sect. 1.3) is another branch of the dispersion equation for the spectrum of emitted photons. In the case of relativistic particles, the diffraction radiation is emitted at a small angle to the particle velocity and its frequency is proportional to $E^2$, contrary to FDPXR, which is independent of the particle energy.

Despite earlier theoretical predictions of FDPXR [11, 13, 16, 21], the first observations have been reported only recently [4, 7, 9] because of the necessity of the HRPXR set-up. Figure 6.6 shows the sketch of the experiment [9].

The general PXR amplitude is calculated from (3.14) by selecting a wave vector $\boldsymbol{k}$, directed to the observation point (see Sects. 2.1 and 3.1). The spectrum of FDPXR (3.14) in a crystal of thickness $L$ follows from (3.19) using a substitution

$$\boldsymbol{e}_{gs} \to \boldsymbol{e}_{s}, \quad \boldsymbol{k}_{-\boldsymbol{g}} \to \boldsymbol{k}, \quad \gamma_{\mu s}^{g} \to \gamma_{\mu s}^{0}, \quad \boldsymbol{v}_{g} \to \boldsymbol{v} .$$

For relativistic electrons with $\gamma = E/mc^2 \gg 1$, for which the CBS contribution is negligible and radiation angle $\theta \approx \gamma^{-1} \ll 1$, the spectral–angular distribution of photons with polarization $\boldsymbol{e}_{\mathrm{s}}$ and per one electron is

$$\frac{\partial^2 N_{gs}}{\partial\omega \partial\boldsymbol{n}} = \frac{\alpha}{\omega\pi^2 c^2} (\boldsymbol{e}_s \boldsymbol{v})^2 \left| \sum_{\mu=1}^{2} \gamma_{\mu s}^{0} \left( \frac{1}{\theta_\gamma^2} - \frac{1}{\theta_{\mu s}^2} \right) M_{\mu s}(L) \right|^2 ;$$

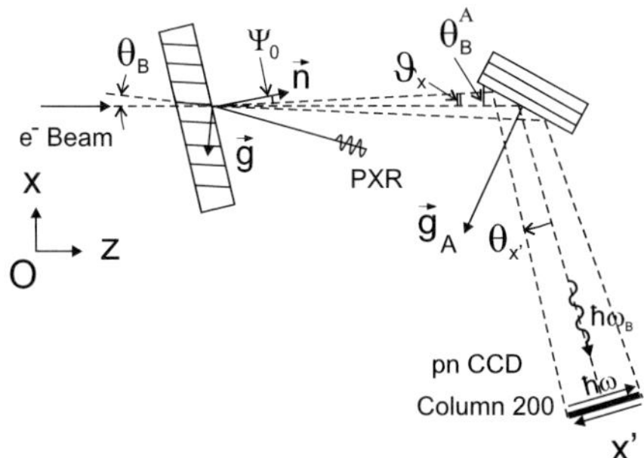


**Fig. 6.6.** Schematic representation of the experimental set-up as the top view from [9]. FDPXR is produced by 855 MeV electrons on the $(11\bar{1})$ planes of Si single-crystal targets of $56, 58, 1,000\,\mu$m thicknesses. Measurements were performed at target Bragg angles $\theta_\mathrm{B} = 9.059°$ and $\theta_\mathrm{B} = 10.797°$corresponding to a PXR photon energy $\hbar\omega_\mathrm{B} = 12.556$ keV and 10.554 keV respectively. The analyser is a Si single crystal located at a distance of 5.5 m or 7.345 m from the target. Forward emitted radiation is reflected at the (333) planes at a Bragg angle $\theta_\mathrm{B}^A = 28.219°$ and detected with the position sensitive pn-CCD detector at a distance 5.5 m from the analyser crystal. The energy of the reflected radiation is a function of the emission angle $\theta_x$

$$
\begin{aligned}
M_{\mu s}(L) &= (\mathrm{e}^{-\mathrm{i}\theta_{\mu s}^2 \omega L/2c\gamma_0} - 1) \; ; \\
\theta_\gamma^2 &= \theta^2 + \gamma^{-2}; \;\; \theta_{\mu s}^2 = \theta^2 + \gamma^{-2} - 2\epsilon_{\mu s} \; ,
\end{aligned}
\tag{6.11}
$$

where $\boldsymbol{e}_1$ is perpendicular to the diffraction plane $\boldsymbol{k}, \boldsymbol{g}$, and $\boldsymbol{e}_2$ lies in this plane.

Because of the coupling of direct and diffracted waves in FDPXR, the geometry (Bragg or Laue) of the experiment influences the FDPXR intensity. For Laue geometry, used in the experiments [7, 9], the coefficients $\gamma_{\mu s}^0$ are

$$
\gamma_{1(2)s}^0 = \frac{2\epsilon_{2(1)s} - \chi_0}{2(\epsilon_{2(1)s} - \epsilon_{1(2)s})} \; . \tag{6.12}
$$

To calculate the radiation amplitude, the vector potential (3.14) with continuous boundary conditions on the crystal surfaces has to be used. Therefore, the spectrum (6.11) includes the radiations due to the jump of dielectric permittivity at both the entrance and exit surfaces of the crystal, the so-called transition X-ray radiation (TXR). To use this classification, we transform (6.11) into

$$
\begin{aligned}
\frac{\partial^2 N_{gs}}{\partial\omega\partial\boldsymbol{n}} &= I_\mathrm{TXR}^{(s)} + I_\mathrm{FDPXR}^{(s)} \; , \\
I_\mathrm{TXR}^{(s)} &= \frac{\alpha p_\mathrm{s}^2 \theta^2}{\omega\pi^2} \frac{|\chi_0|^2}{\theta_\gamma^4 |\theta_\chi^2|^2} \left| \sum_{\mu=1}^{2} \gamma_{\mu s}^0 M_{\mu s}(L) \right|^2 \; ,
\end{aligned}
$$

$$I^{(s)}_{\rm FDPXR} = \frac{\alpha p_{\rm s}^2\theta^2}{\omega\pi^2}\left\{\left|\sum_{\mu=1}^{2}\gamma^0_{\mu s}\left(\frac{1}{\theta^2_\chi}-\frac{1}{\theta^2_{\mu s}}\right)M_{\mu s}(L)\right|^2 - 2\sum_{\mu=1}^{2}\sum_{\nu=1}^{2}{\rm Re}\left[\gamma^0_{\mu s}\gamma^{0*}_{\nu s}M_{\mu s}(L)M^*_{\nu s}(L)\left(\frac{1}{\theta^2_\chi}-\frac{1}{\theta^2_{\mu s}}\right)\left(\frac{1}{\theta^2_\chi}-\frac{1}{\theta^2_{\gamma}}\right)^*\right]\right\},$$

$$\theta^2_\chi = \gamma^{-2}+\theta^2-\chi_0, \qquad p_1=\sin\varphi, \qquad p_2=\cos\varphi\,, \tag{6.13}$$

where $\varphi$ is an angle between the diffraction plane and radiation plane $(\boldsymbol{k},\boldsymbol{v})$.

The first term in (6.13) differs from the regular TXR spectral–angular distribution in a homogeneous plate of thickness $L$ [26] by the dependence of interference of radiation at the entrance and exit crystal surfaces on the effective refraction index $(1+2\epsilon_{\mu s})$. This effect was observed in [9] and used for weakening of the TXR contribution by selection of a proper crystal thickness.

The principal obstacle for experimental observation of FDPXR is the concentration of latter within the narrow angular cone $\sim\gamma^{-1}$ around the particle velocity, where the photons produced by any radiative mechanism are transmitted. This background is absent in the case of wide-angle PXR reflections. To analyse the observation conditions for FDPXR, the signal/noise ratio is derived below [16] from (6.13). Let us assume that the photons of frequency $\omega$ are detected in the angular region $\theta\le\gamma^{-1}$ by the detector with the energy resolution $\hbar\Delta\omega_{\rm D}$ and aperture $\Delta\Omega_{\rm D}\le\gamma^{-2}$. Then, the estimate for the first term in (6.13) follows straightforwardly, which does not depend on the crystal thickness (interference and polarization effects are neglected):

$$\Delta N_{\rm TXR}\approx\frac{\alpha}{\pi^2}\frac{|\chi_0|^2\gamma^6}{(1+\gamma^2|\chi_0|)^2}\frac{\Delta\omega_{\rm D}}{\omega}\Delta\Omega_{\rm D}\,. \tag{6.14}$$

For FDPXR photons, the maximal spectral density of PXR corresponds to the Cherenkov peak, which is conditioned by

$${\rm Re}(\theta^2_{\mu s}) = {\rm Re}(\theta^2+\gamma^{-2}-2\epsilon_{\mu s}) = 0\,. \tag{6.15}$$

For thick crystals, $4\omega\,{\rm Im}(\epsilon_{\mu s})L\gg1$, (6.13) (see also Sect. 3.4) gives a maximal peak intensity:

$$I^{\rm max}_{\rm FDPXR}\sim\frac{1}{4|{\rm Im}(\epsilon_{\mu s})|^2}, \qquad \Delta\alpha_{\rm B}\approx\frac{\Delta\omega_{\rm B}}{\omega_{\rm B}}\approx 2|{\rm Im}(\epsilon_{\mu s})|\,, \tag{6.16}$$

where the deviation from the Bragg condition, $\Delta\alpha_{\rm B}$, defines the spectral width $\Delta\omega_{\rm B}$ of the peak, according to (3.26). Using formula (3.15) for the effective polarizability $\epsilon_{\mu s}$ and condition $\Delta\omega_{\rm B}\ll\Delta\omega_{\rm D}$, which is satisfied for every detector, the integral number of photons in the FDPXR peak is

$$\Delta N_{\rm FDPXR}\approx\frac{\alpha}{\pi^2}\left|\frac{|\chi_g|^2\gamma^4}{(1+\gamma^2|\chi_0|)^2+|\chi_g|^2\gamma^4}\right|^2\frac{1}{\gamma^2|{\rm Im}(\chi_0)|}\Delta\Omega_{\rm D}\,. \tag{6.17}$$

Thus, the signal/noise ratio in the FDPXR experiment is expected to be

$$\xi_F \approx \left| \frac{|\chi_g|^2(1+\gamma^2|\chi_0|)}{|\chi_0|\left[(1+\gamma^2|\chi_0|)^2+|\chi_g|^2\gamma^4\right]} \right|^2 \frac{1}{|\mathrm{Im}(\chi_0)|} \frac{\omega}{\Delta\omega_\mathrm{D}} \,. \tag{6.18}$$

The ratio $\xi_F$ depends on the particle energy and tends to 1 if the energy resolution of the detector is comparable with the width of the Bragg reflection from the analyser. For the optimal particle energy $\gamma^2|\chi_0| \approx 1$ and detector resolution $\Delta\omega_\mathrm{D} \approx |\chi_g|\omega$:

$$\xi_\mathrm{opt} \approx \frac{1}{4}\frac{|\chi_g|^2}{|\chi_0|^2}\frac{|\chi_g|}{|\mathrm{Im}\,(\chi_0)|} \,. \tag{6.19}$$

In general, this estimate must include the contributions of other radiation types, e.g. bremsstrahlung. However, assuming a narrow spectral interval of the detector, these contributions can be neglected (see (4.24)). Substituting a typical value of X-ray polarizability [1] into (6.19), the value $\xi_\mathrm{opt}$ is found to be varied within the limits 0.1–1.0. In successful FDPXR experiments [4,9] with a double-crystal set-up, the low peak magnitude fits the estimate (6.19) well. There are several published proposals for improvement of the signal/noise ratio in FDPXR experiments. In [16], two set-ups have been suggested: (i) use of difference between parametric and transition radiations in the polarization plane; (ii) use a spectra subtraction at different positions of the target crystal. The different behaviour of the dependence [11] of radiation spectra on the particle energy for PXR and transition radiation allows us to distinguish between them. The first FDPXR experiment [7] used the precise measurement of orientational dependence of TXR from entrance and exit surfaces of a thin target crystal. Figure 6.7 from [7] demonstrates the explicit peak formed by constructive interference of TXR from both surfaces. This peak is described by the first term in (6.13) and is conditioned by $\theta^2_{\mu s}\omega L/2c\gamma_0 = \pi$. The diffraction condition (6.10) has to be satisfied to change the position of the peak for small variation of radiation angle (Fig. 6.7). The opposite case of destructive interference is shown in Fig. 6.8a from [9], where the thickness of the crystal, $L_1$, was chosen to satisfy the condition $\theta^2_{\mu s}\omega L_1/2c\gamma_0 = 2\pi$.

Figure 6.8b shows the direct FDPXR measurements in [9] from the thick crystal, when the background is caused by TXR from the exit surface. The ratio of the peak intensity to background is also well explained by (6.19).

## 6.3 Multiwave PXR

The multiwave diffraction of emitted parametric X-rays in a crystal gives one more proof of PXR dynamical nature in thick perfect crystals. Both classic theory in Sect. 2.1 and quantum theory in Sect. 3.1 derive the PXR intensity from homogeneous Maxwell's equations and boundary conditions, which also

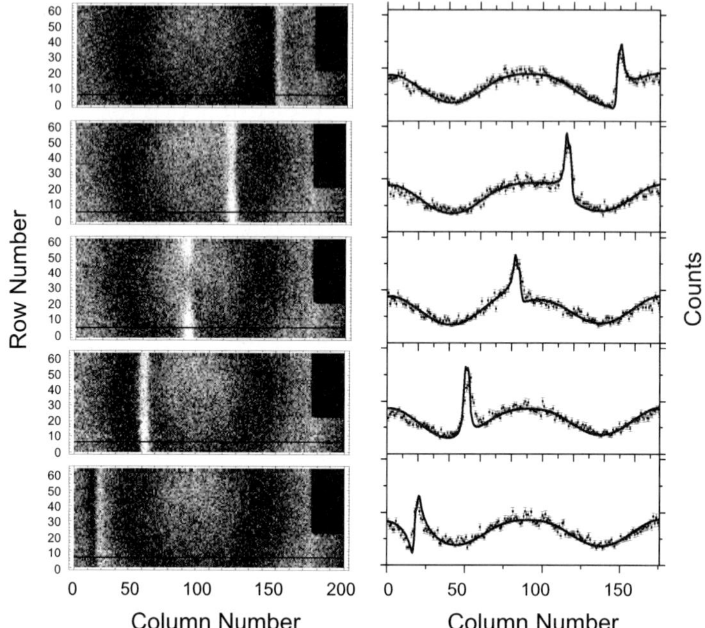


**Fig. 6.7.** Results of the experiment [7] for the photon energy of 12.556 keV and the target thickness of 56 μm. The white vertical lines in the left panels are constructive interference fringes as taken by the pn-CCD detector. From the upper to the lower panel the rotation angle $\psi_x$ of the target crystal increases, starting from 0.0007 rad, in steps $\Delta\psi_x = -0.00035$ rad. The error bars in the right panels show projected spectra of the pn-CCD detector rows. The full lines represent calculations on the basis of formulas from [5], which are analogous to (6.13) yet with other notations

describe the conventional X-ray diffraction in the crystal from an external X-ray source. The difference between the expressions for electromagnetic fields for both processes, PXR and X-ray diffraction, is only in the choice of the wave vector $\boldsymbol{k}$ (see Appendix A.2). Therefore, the known features of multiwave HRXRD [14] have to have the analogues in HRPXR.

The detailed consideration of multiwave PXR has been done in [15, 25], and here we present a general approach to theoretical treatment of this case. Assume that the wave vector satisfies the Bragg condition for $N$ reciprocal lattice vectors simultaneously ($N$ is the order of multiple Bragg diffraction):

$$\begin{aligned} \boldsymbol{k}_m &= \boldsymbol{k} + \boldsymbol{m}, \qquad \boldsymbol{m} = 0, \boldsymbol{g}, \boldsymbol{h}, \dots , \\ \alpha_m &= \frac{k_m^2 - k^2}{k^2} \leq |\chi_m|, \qquad m = 1, 2, \dots, N . \end{aligned} \tag{6.20}$$

In this case the wave field $\boldsymbol{E}_{\boldsymbol{k}}^{(s)}(\boldsymbol{r}, \omega)$ inside a crystal can be represented as a sum of transverse Bloch waves with polarizations $\boldsymbol{e}_m^{(s)}$, $s = \sigma, \pi$, which travel

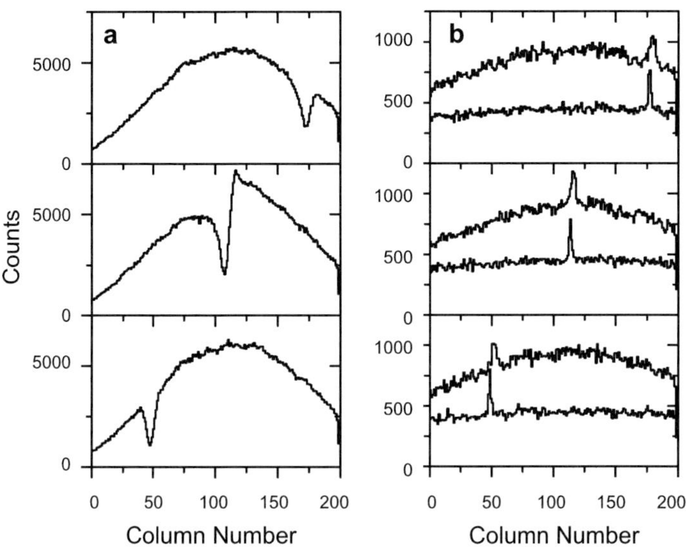


**Fig. 6.8.** Results of the experiment [19] for the photon energy of 10.554 keV and the target thickness of 58 μm (**a**) and 1,000 μm (**b**). The intensity distributions are summed over all 64 rows of the pn-CCD detector as a function of the column number. From the upper to the lower panel, the rotation angle $\psi_x$ of the target crystal decreases starting from 0.0005435 rad, in steps $\Delta\psi_x = -0.0005894$ rad. Destructive interference can clearly be recognized in (a). Panel (b), upper curves: beam spot size about 560 × 434 μm$^2$, the pn-CCD detector was shifted to the observation angle $\theta_y = 0.0006$ rad. Panel (b), lower curves: reduced beam spot size about 114 × 200 μm$^2$, the direction of the electron beam was altered resulting in an effective vertical observation angle $\theta_y = 0.0004$ rad and a shift of 2.6 pixels horizontally

in the direction of Bragg diffraction $\boldsymbol{k}_m$ and correspond to $2N$-dispersion branches [14]:

$$\boldsymbol{E}_{\boldsymbol{k}}^{(s)}(\boldsymbol{r},\omega) = \sum_m \boldsymbol{e}_m^{(s)} B_m^{(s)}(\boldsymbol{r}) \, ,$$

$$B_m^{(s)}(\boldsymbol{r}) = \sum_{j=1}^{N} \lambda^{(j)} D_m^{(s,j)} \exp(\mathrm{i}[\boldsymbol{k}_m - (\epsilon^{(j)} - \alpha_m/2\gamma_m)\boldsymbol{n}\omega/c]\boldsymbol{r}) \, . \quad (6.21)$$

Here $\gamma_m$ are the cosines of the angles between the X-ray wave vector and the internal normal to the crystal surface, the parameters $\lambda^{(j)}$ are the coefficients of excitations of various dispersion branches with amplitudes $D_m^{(s,j)}$ and $\epsilon^{(j)}$ defines the refraction of the corresponding Bloch waves depending on the deviation parameters $\alpha_m$.

The wave amplitudes $D_m^{(s,j)}$ and the values $\epsilon^{(j)}$ can be found as the eigenvectors and eigenvalues of the dynamical diffraction equations which have the form of $2N \times 2N$ matrix equations [14]:

$$\sum_{s'}\sum_{m'} G_{mm'}^{ss'} D_m^{(s',j)} = 2\epsilon^{(j)} D_m^{(s,j)} ,$$

$$G_{mm'}^{ss'} \equiv \alpha_m \delta_{mm'}^{ss'} - \frac{1}{\gamma_m}\Big\{ \chi_{m-m'}(\boldsymbol{e}_m^{(s)}\boldsymbol{e}_{m'}^{(s')}) + \mathrm{i}\chi_{m-m'}^{Q}/k^2 \Big[(\boldsymbol{k}_m\boldsymbol{k}_{m'})(\boldsymbol{e}_m^{(s)}\boldsymbol{e}_{m'}^{(s')}) + (\boldsymbol{k}_m\boldsymbol{e}_{m'}^{(s')})(\boldsymbol{e}_m^{(s)}\boldsymbol{k}_{m'})\Big]\Big\} , \quad (6.22)$$

where $\chi_{m-m'}, \chi_{m-m'}^{Q}$ are the dipole and quadrupole components of the crystal dielectric polarizability $\chi(\boldsymbol{r},\omega)$ in a Fourier series over the reciprocal lattice vectors.

The coefficients $\lambda^{(j)}$ are defined due to the boundary conditions. For a plate-shaped crystal these conditions are reduced to the system of linear equations

$$\sum_{j=1}^{2N} C_m^{(s,j)}\lambda^{(j)} = \delta_{m0}(\delta^{s\sigma}\cos\varphi + \delta^{s\pi}\sin\varphi) , \quad (6.23)$$

with $C_m^{(s,j)} = D_m^{(s,j)}$ for the Laue-case Bragg waves ($\gamma_m > 0$) and $C_m^{(s,j)} = D_m^{(s,j)}\exp(-\mathrm{i}\epsilon^{(j)}\omega L/c)$ for the Bragg case ($\gamma_m < 0$); $L$ is the thickness of the crystal plate and $\varphi$ is the angle between the polarization plane of the incident wave defined by the detector and vector $\boldsymbol{e}_m^{(s)}$.

Substituting (6.21)–(6.23) into the general formula (2.16), we find the general expression for the spectral–angular distribution of multiwave PXR. Let us write it only for the case of the thick crystal:

$$\frac{\partial^2 N_{gs}}{\partial\omega\partial\boldsymbol{n}} = \frac{\alpha}{\omega_{\mathrm{B}}\pi^2}\left|\sum_{s'}\sum_{j=1}^{2N}(\boldsymbol{v}\boldsymbol{e}_g^{(s')})\lambda^{(j)}(\omega)\frac{D_g^{(s',j)}(\omega)}{Q^{(j)}(\omega)}\right|^2 ,$$

$$Q^{(j)}(\omega) = \gamma^{-2} + \theta_x^2 + \theta_y^2 + \theta_{\mathrm{s}}^2 + \gamma_g(2\epsilon^{(j)} - \alpha_m) . \quad (6.24)$$

This expression is a generalization of the two-beam dynamical diffraction formula (3.27) for the multiwave case. As in (3.27), the frequency $\omega_{\mathrm{B}}$ corresponds to the exact Bragg condition and radiation angles $\theta_x, \theta_y$ are chosen in the plane perpendicular to $\omega_{\mathrm{B}}\boldsymbol{v}/c^2 + \boldsymbol{g}$.

There are some new qualitative PXR features expected for multiwave diffraction. First of all, the forbidden reflection can be excited, for example, the (222) PXR reflection in a germanium crystal [15]. This becomes possible due to coupling in a three-wave Bragg diffraction of two planes $(3\bar{1}\bar{1})$ and $(\bar{1}33)$, analogously to the Renninger effect in conventional X-ray diffraction. The angular distribution in the (222) reflection differs essentially from the two-beam case, because the radiation probability is localized in the narrow spectral domain. This effect is not yet experimentally investigated.

The second important feature of multiwave PXR, which was also considered in [15, 25], is a fine structure of angular distribution and gain of the

Cherenkov peak intensity inside a narrow spectral interval. The PXR distribution inside reflection is determined by two factors in (6.24): the amplitude $D_g^{(s',j)}(\omega)$, which is close to unity near the Bragg condition:

$$|\alpha_g| < |\chi_g|, \qquad |D_g^{(s',j)}(\omega)| \approx 1 \,, \tag{6.25}$$

and the parameter $Q^{(j)}(\omega)$, which defines the position of the Cherenkov peak:

$$\mathrm{Re}[Q^{(j)}(\omega)] = \gamma^{-2} + \theta_x^2 + \theta_y^2 + \theta_\mathrm{s}^2 + \gamma_g[2\mathrm{Re}(\epsilon^{(j)}) - \alpha_m] = 0 \,. \tag{6.26}$$

If both conditions (6.25) and (6.26) are fulfilled, the PXR intensity gets a gain:

$$\xi \approx \left| \frac{\chi |g|}{\mathrm{Im}\,(\chi_0)} \right|^2 \sim 10^2\text{–}10^3 \,.$$

In the case of two-beam diffraction (see Sects. 3.4 and 6.1), the Cherenkov peak is conditioned by (6.26) but its position leaves out the region (6.25), which suppresses the peak amplitude by $|\chi_g|^2/|\chi_0|^2$ times. In the multiwave case, conditions (6.25) and (6.26) are fulfilled simultaneously, which results in fine structure of PXR reflection [15, 25], which is analogous to Kössel lines in conventional multiwave diffraction [14]. The detailed experimental investigation of multiwave effects requires a HRPXR technique because the spectral–angular interval, where they become apparent, is considerably less than the reflection width.

The first experiments on angular distribution of PXR reflection in the multiwave case were carried out at synchrotron Sirius (Tomsk, Russia) [2, 3].

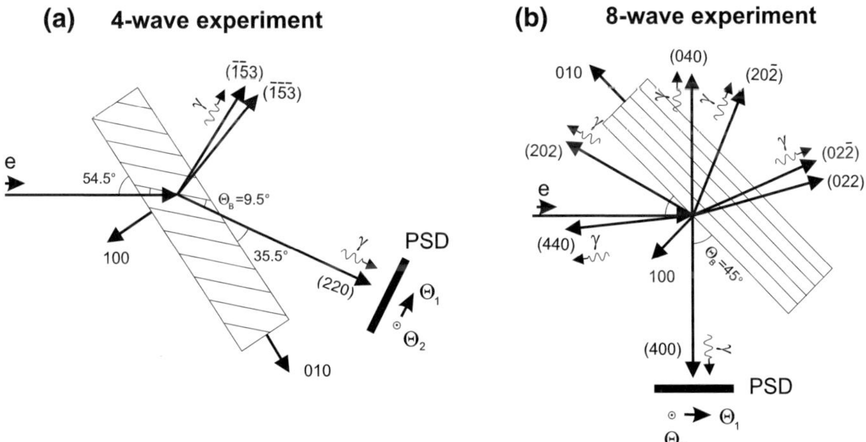


**Fig. 6.9.** Schematic layouts of four-wave (**a**) and eight-wave (**b**) PXR experiments. All coupled Bragg reflections are shown. The angles $\theta_x$ and $\theta_y$ are varied along the PSD rule and perpendicularly to the picture plane. $\theta_\mathrm{B}$ indicates the Bragg angle for the measured reflection

The electron beam of 500 MeV (four-wave experiment) and 900 MeV (eight-wave experiment) has been used. The schematic layout of the experiments is shown in Fig. 6.9. A 400-μm-thick GaAs single-crystal plate cut parallel to the (100) plane was set on the goniometer in such a way that the electron beam moved through it nearly along the incidence direction for the four-beam Bragg diffraction $(000), (220), (\overline{153}), (\overline{1}5\overline{3})$ of X-rays with $\lambda = 0.6708$ Å in the experiment [3], and for the eight-beam diffraction $(000), (400), (022), (02\overline{2}), (202), (20\overline{2}), (040), (440)$ of X-rays with $\lambda = 1.9987$ Å in [2]. In the latter case, the incidence direction was parallel to the $\langle 110 \rangle$ crystallographic axis. The experiments were aimed to reveal the changes in one of the PXR reflections due to coupling with X-ray beams radiated along the other directions. Therefore, only one PXR beam corresponding to the Bragg planes (220) and (400) was registered in [3] and [2], respectively. The radiation was detected by a linear position sensitive detector (PSD). A series of PXR angular distributions as functions of $\theta_x$ on the PSD rule were measured while varying the angle $\theta_y$ by a linear displacement of the detector (Fig. 6.9). The PXR angular distributions measured in the experiments [2,3] are shown in Figs. 6.10 and 6.11, respectively.

The PXR angular distribution corresponding to the four-wave diffraction geometry [3] displays an asymmetry in the height of the right and left peaks. This is contrary to the case of two-wave PXR generation, where the peaks are symmetrical. The theoretical interpretation of these peaks [25] displayed the contribution of both multiwave effects and non-homogeneous beam cross-section to the asymmetry of the curve (see also the analysis of the asymmetry of the angular distribution in Sect. 5.4).

In the case of eight-wave PXR generation (Fig. 6.11), the prominent asymmetry of two peaks (the left peak is 2.5 times higher than the right one) and the splitting of the left peak into two parts is seen. For comparison, the respective

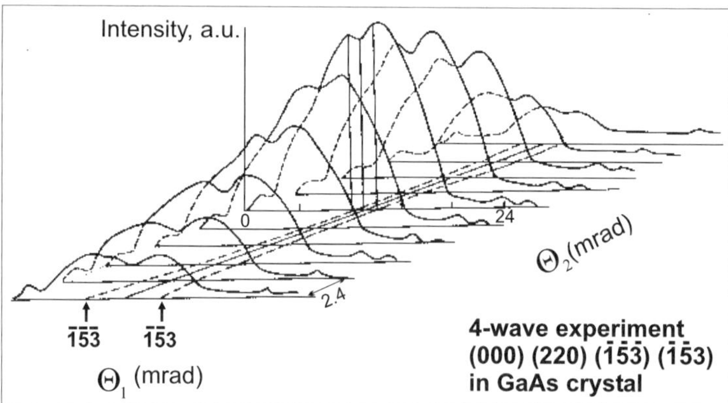


**Fig. 6.10.** The PXR angular distribution for the (220) reflection in GaAs measured in [3] under the four-wave diffraction condition

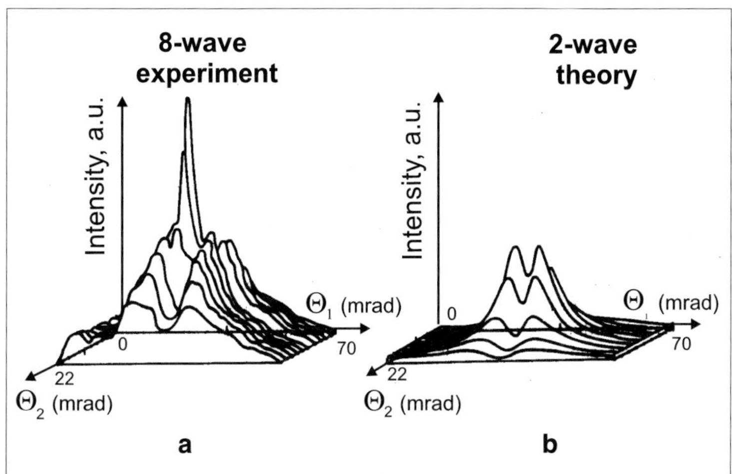


**Fig. 6.11.** The PXR angular distribution for the (400) reflection in GaAs under the eight-wave diffraction condition: (**a**) the experiment [2]; (**b**) the theoretical distribution calculated in two-wave approximation

two-wave angular distribution calculated under the same experimental conditions is presented in Fig. 6.11b. The theoretical analysis of this experiment [25] demonstrated the extension of the spectral–angular interval, where the cross-influence of different reflections is essential. This fact explains qualitatively the experiment [2]; however, quantitative fitting requires the consideration of instrumental effects.

The direct confirmation of multiwave PXR has been also recently obtained in [7]. The angular distribution of the PXR reflection from the (111) Si crystal has been measured in backward diffraction geometry. The experimental set-up was similar to Fig. 6.1, however, the pn CCD detector was used to register separately two-dimensional angular distributions for different reflections $(nnn)$, $n = 1, 3, 4, 5 \ldots$. The condition for observation of multiwave PXR in this geometry is the equality of the reflection frequency $\omega_{\mathrm{B}}^{(n)} = cg_n/2$ and the Bragg frequency for other selected reflection $(hkl)$ with properly chosen indices. Using (2.34), this condition is analytically written as [7]

$$\omega_{hkl} = \frac{cg_{hkl}}{2\cos\theta_0}, \quad \cos\theta_0 = \frac{(\boldsymbol{g}_{hkl}\boldsymbol{g}_n)}{g_{hkl}g_n} = \frac{h+k+l}{\sqrt{3(h^2+k^2+l^2)}}\,;$$
$$n = \frac{h^2+k^2+l^2}{h+k+l}\,. \tag{6.27}$$

This equation is satisfied for $n = 3, 4, 5$ for the reflections $(11\bar{1})$ and (224), (004) and (044), $(\bar{1}35)$ and (026), respectively.

Typical intensity distributions are shown in Fig. 6.12 from the paper [7]. The predicted multiwave interference appears indeed for the (333) and (555)

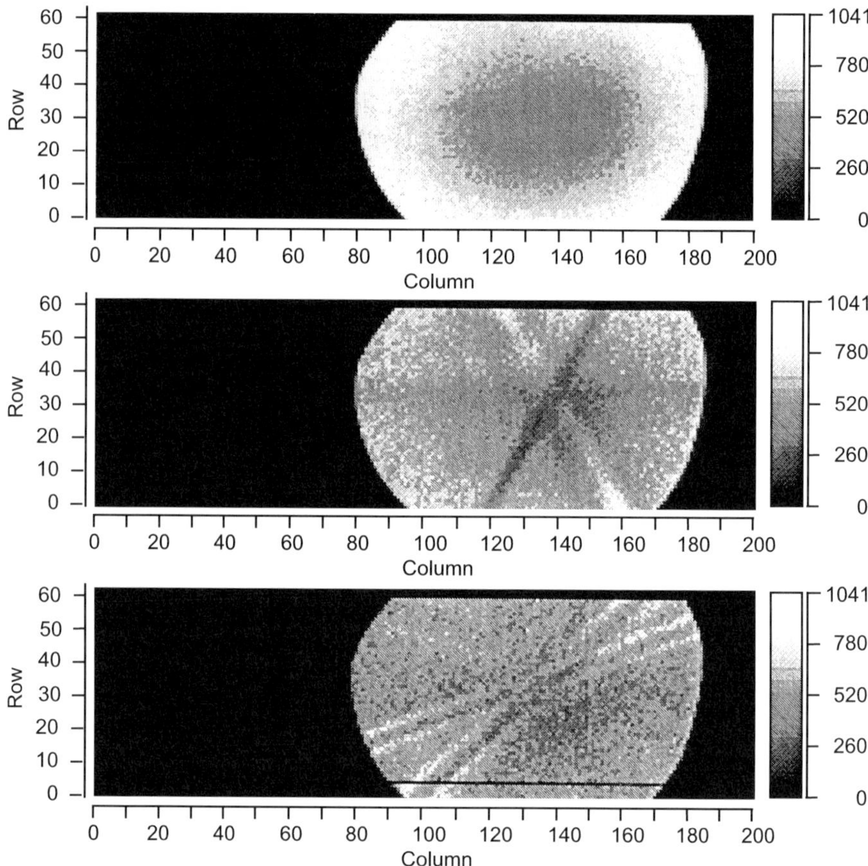


**Fig. 6.12.** Intensity distributions in the backward direction taken with the pn-CCD detector [7]. In the upper, middle and lower panels, intensity distributions for the (111), (333) and (555) reflections are shown, respectively. A vacuum tube with an inner diameter of 20 nm limited the view area

reflections, while the (111) reflection shows the typical smooth distribution. The marked regions on the diagrams for the (333) and (555) reflections fit well the theoretical prediction for Kössel line positions in four-wave PXR [7]. The asymmetry of the (333) reflection, however, is not explained by theory and caused probably by the deformation of the crystal surface [7].

## 6.4 PXR in the Degenerate Case of Two-Beam Diffraction

Another manifestation of dynamical diffraction in PXR has been predicted in [18] and not yet been experimentally observed. It is known [23] that the conventional diffraction problem has a singularity when two X-ray eigenwaves are degenerated, i.e. their refraction indices are equal to each other. In this case, the wave fields are described by so-called Voigt waves instead of superposition of plane waves under usual conditions. This singularity should affect the PXR spectrum for certain orientation of the crystallographic planes relative to the electron velocity.

The above-mentioned degeneracy exists in the case of the Bragg diffraction geometry only, when the wave field is described by (3.14)–(3.16). Then the general expression (3.19) for the spectral–angular distribution of PXR quanta radiated per relativistic electron has the following form:

$$\begin{aligned}\frac{\partial^2 N_{gs}}{\partial\omega\partial\, \boldsymbol{n}} &= \frac{\alpha|\chi_g^{(s)}|^2}{4\pi^2c^2\omega}\theta^2 p_{\mathrm{s}}^2\beta^2\left|\frac{1-\mathrm{e}^{\mathrm{i}q_1^{(s)}\omega L}}{q_1^{(s)}}-\frac{1-\mathrm{e}^{\mathrm{i}q_2^{(s)}\omega L}}{q_2^{(s)}}\right|^2\\ &\times\left|(2\epsilon_2^{(s)}-\chi_0)\exp\left(\mathrm{i}\frac{\omega}{\gamma_0}\epsilon_1^{(s)}L\right)-(2\epsilon_1^{(s)}-\chi_0)\exp\left(\mathrm{i}\frac{\omega}{\gamma_0}\epsilon_2^{(s)}L\right)\right|^{-2},\\ q_{1,2}^{(s)} &= \frac{1}{2}\left(\theta^2+\gamma^{-2}-2\epsilon_{1,2}^{(s)}\right).\end{aligned} \tag{6.28}$$

The notations for (6.28) are introduced in (3.14)–(3.16), and the polarization factor $p_{\mathrm{s}}$ is defined in (6.13). The high-energy electrons are considered here to skip multiple scattering and CBS effects.

When the equality $\epsilon_1^{(s)}=\epsilon_2^{(s)}$ is fulfilled (degenerate diffraction case), both factors in (6.28) are equal to zero and the detailed analysis of the uncertainty is necessary near the singularity. Let us introduce the definition

$$\chi_g^{(s)}\chi_{-g}^{(s)}\equiv z_{\mathrm{s}}'+\mathrm{i}z_{\mathrm{s}}'', \qquad \chi_0=\chi_0'+\mathrm{i}\chi_0''$$

and consider the equations which define the degenerate case for PXR. The first one is the equality of the refraction indices:

$$\mathrm{Re}(\epsilon_1^{(s)}-\epsilon_2^{(s)})=0\,. \tag{6.29}$$

The second equation is the Cherenkov condition for PXR:

$$\mathrm{Re}\;q_{1,2}^{(s)}=\frac{1}{2}\left(\theta^2+\gamma^{-2}-2\epsilon_{1,2}^{(s)}\right)=0\,. \tag{6.30}$$

The solution for (6.29), (6.30) defines the resonant values of the radiation angle $\theta_0$ and the deviation parameter $\alpha_{\mathrm{B}}^{(0)}$:

$$\begin{aligned}\theta_0^2 &= (-\beta z_{\mathrm{s}}')^{1/2}-\gamma^{-2}-|\chi_0'|, \qquad |\beta|>\frac{|\chi_0'|^2}{z_{\mathrm{s}}'},\\ \alpha_{\mathrm{B}}^{(0)} &= -\frac{\chi_0'(1-\beta)+2(-\beta z_{\mathrm{s}}')^{1/2}}{\beta}\,.\end{aligned} \tag{6.31}$$

The values $\chi_0'$ and $z_{\mathrm{s}}'$ have to be calculated for the frequency $\omega_{\mathrm{B}}=g/2\sin\theta_{\mathrm{B}}$ corresponding to the centre of PXR reflection.

From (6.31) it follows that the considered singularity takes place only for the case $\beta < 0$, $|\beta| \geq 1$, which corresponds to the asymmetric Bragg diffraction geometry. In order to analyse the PXR intensity, we expand the most essential terms in formula (6.28) into the power series of $(\alpha_{\mathrm{B}} - \alpha_{\mathrm{B}}^{(0)})$:

$$\epsilon_{1,2}^{(s)} = \frac{1}{2}(a \pm \Delta\epsilon_{\mathrm{s}}), \qquad q_{1,2}^{(s)} = q_0 \mp \frac{1}{2}\Delta\epsilon_{\mathrm{s}}, \qquad q_0 = \frac{1}{2}(\gamma^{-2} + \theta^2 - a)\,,$$

$$a = \gamma^{-2} + \theta^2 + \mathrm{i}\chi_0''(1+\beta) - \frac{1}{2}\beta(\alpha_{\mathrm{B}} - \alpha_{\mathrm{B}}^{(0)})\,,$$

$$\Delta\epsilon_{\mathrm{s}} = (-\beta z_{\mathrm{s}}')^{1/2}\left[-\beta(\alpha_{\mathrm{B}} - \alpha_{\mathrm{B}}^{(0)}) - \mathrm{i}\chi_0''(1 - \beta + 2\frac{\mathrm{Im}\ \chi_g^{(s)}}{\chi_0''}\sqrt{|\beta|})\right]^{1/2} \tag{6.32}$$

The imaginary part of $\Delta\epsilon_{\mathrm{s}}$ is non-zero for any $\alpha_{\mathrm{B}}$ because of the conditions $\beta < 0$ and $\chi_0'' > \mathrm{Im}\ \chi_g^{(s)}$. Thus, for a sufficiently thick crystal the inequality

$$\exp\left[\mathrm{i}\omega(\epsilon_1^{(s)} - \epsilon_2^{(s)})\frac{L}{c}\right] \gg 1$$

is always fulfilled, and there is no singularity in the denominator of formula (6.28). Therefore, let us consider a crystal with thickness $L$, which is larger than the primary extinction length $L_{\mathrm{ext}}$ for the radiation, but smaller than the absorption length in the degenerate case, until the PXR spectral intensity includes the singularity, that is

$$L_{\mathrm{ext}} = \frac{c}{\omega|\chi_0'|} \ll L \ll \frac{c}{\omega \mathrm{Im}\ \Delta\epsilon_{\mathrm{s}}} < \frac{c}{\omega|\chi_0'|} = L_{\mathrm{abs}}\,. \tag{6.33}$$

The inequalities (6.33) permit us to transform formula (6.28) in the vicinity of its maximum as follows:

$$\frac{\partial^2 N_{gs}}{\partial\omega\partial\,\boldsymbol{n}} = \frac{\alpha|\chi_g^{(s)}|^2}{4\pi^2c^2\omega}\theta^2 p_{\mathrm{s}}^2\beta^2\frac{\sin^2\kappa\xi/4}{(\mu\xi/4)^2} \times \frac{(1+\tan^2\kappa)}{[(\theta^2 + \gamma^{-2} + |\chi_0'|)\tan^2\kappa + 4\mu^2]^2}\,, \tag{6.34}$$

where

$$\mu = (-\beta z_{\mathrm{s}}')^{1/2}, \qquad \xi = -\frac{\beta(\alpha_{\mathrm{B}} - \alpha_{\mathrm{B}}^{(0)})}{4z_{\mathrm{s}}'} - 1, \qquad \kappa = \mu\omega L/c\,.$$

The PXR spectral intensity (6.34) in the case of degenerate diffraction is the oscillating function of the parameter

$$\kappa = (-\beta z_{\mathrm{s}}')^{1/2}\omega L/c\,.$$

These oscillations take place at the variation of both the crystal length and the orientation of the crystal relative to the electron velocity. Even the total

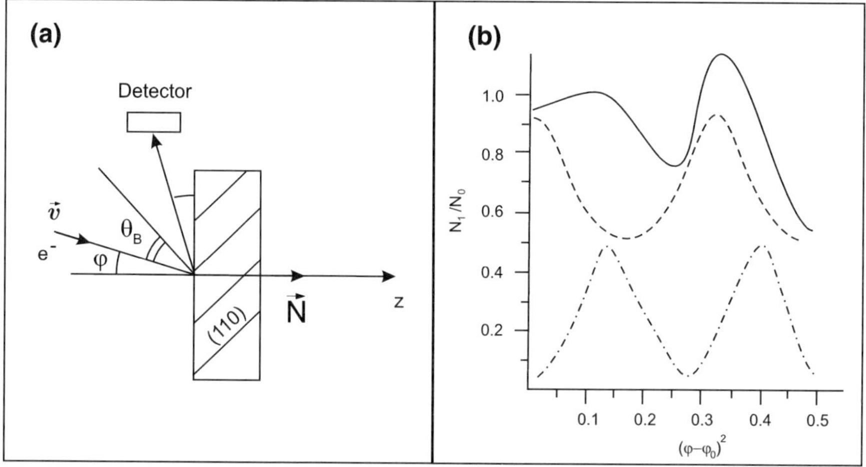


**Fig. 6.13.** Oscillations of the PXR intensity of the (220) reflection in diamond as a function of the electron incidence angle: (**a**) schematic experimental layout; (**b**) theoretical values are calculated for the parameters $\varphi_0 = 5°$, $\theta_D = 2.5°$, $E =$ 900 MeV, $L = 0.01$ cm, $N_0 = 3.5 \times 10^{-6}$ quanta/electron (*dotted*, *dashed* and *solid* curves correspond to $\sigma$-polarization, $\pi$-polarization and the total number of emitted quanta, respectively)

number of quanta registered by the detector with angular width $\theta_D \gg 1$ oscillates as well. This number is calculated by integrating formula (6.34):

$$N_{gs} = \frac{\alpha|\chi_g^{(s)}|^2}{16\sin^2\theta_B}|\beta|\frac{\omega_B L}{c}\left(1 + \frac{1}{B^2}\right) \times \left\{\ln\frac{(1+\rho_D)^2 + A^2}{1 + A^2} - \frac{1}{A}\left(\tan^{-1}\frac{1+\rho_D}{A} + \tan^{-1}\frac{1}{A}\right)\right\}, \quad (6.35)$$

where

$$A = \frac{2(-\beta z_s')^{1/2}}{B(\gamma^{-2} + |\chi_0'|)}, \quad B = \tan\left[(-\beta z_s')^{1/2}\omega_B L/c\right], \quad \rho_D = \frac{\theta_D^2}{(\gamma^{-2} + |\chi_0'|)}.$$

As an example, Fig. 6.13 shows the geometry of the possible experiment for degenerate diffraction observation and the number of PXR photons as a function of the angle between the electron beam velocity and the crystallographic planes. The oscillations of the value $N_s$ are different for both polarizations and can be observed under conditions of HRPXR, because they are defined by sufficiently small variation of the crystal orientation.

## References

1. Yu.N. Adishchev, V.A. Verzilov, A.P. Potylitsyn, S.R. Uglov, S.A. Vorobyev: Nucl. Instrum. Methods B **44**, 130 (1989)

2. V.P. Afanasenko, V.G. Baryshevsky, S.V. Gatsicha, R.F. Zuevsky, M.G. Lifshits, A.S. Lobko, V.I. Moroz, V.V. Panov, I.V. Polikarpov, V.P. Potsiluiko, P.F. Safronov, A.O. Yurtsev: Sov. JETP Lett. **51**, 213 (1990)
3. V.P. Afanasenko, V.G. Baryshevsky, O.T. Gradovsky, R.F. Zuevsky, M.G. Livshits, A.S. Lobko, V.I. Moroz, V.V. Panov, I.V. Polikarpov, D.S. Shvarkov, A.O. Yurtsev: Phys. Lett. A **141**, 311 (1989)
4. A.N. Aleinik, A. Saryshev, E.A. Bogomazova, B.N. Kalinin, G.A. Naumenko, A.P. Potylitsin, G.A. Saruev, A.F. Sharafutdinov: JETP Lett. **79**, 396 (2004)
5. S. Asano, I. Endo, M. Harada, S. Ishii, T. Kobayashi, T. Nagata, M. Muto, K. Yoshida, H. Nitta: Phys. Rev. Lett. **70**, 3247 (1993)
6. A. Authier: *Dynamical Theory of X-ray Diffraction* (Oxford University Press, New York 2001) pp 3–26
7. H. Backe, C. Ay, N. Clawiter, Th. Doerk, M. El-Ghazaly, K.-H. Kaiser, O. Kettig, G. Kube, F. Hagenbuck, W. Lauth, A. Rueda, A. Sharafutdinov, D. Schroff, T. Weber: *Proc. Int. Symp. on Channelling – Bent Crystals – Radiation Processes* (J.W. Goethe University, Frankfurt am Main 2003), p 41
8. H. Backe, G. Kube, W. Lauth: *Proc. NATO Advanced Research Workshop on Electron-Photon Interaction in Dense Media,* ed. by Wiedeman (Kluwer, Dordrecht, Boston, London 2001) NATO Science Series II, vol 49, p 153
9. H. Backe, A. Rueda, W. Lauth, N. Clawiter, M. El-Ghazaly, R. Kunz, T. Weber: Nucl. Instrum. Methods B, **234**, 138 (2005)
10. V.G. Baryshevsky: Dokl. Akad. Sci. Belarus **15**, 306 (1971); V.G. Baryshevsky, I.D. Feranchuk: *Proc. XXI Conf. on Nuclear Structure and Nuclear Spectroscopy* (Moscow University, Moscow 1971) p 220
11. V.G. Baryshevsky: Nucl. Instrum. Methods B **122**, 13 (1997)
12. V.G. Baryshevsky, I.D. Feranchuk: Phys. Lett. A **57**, 183 (1976)
13. A. Caticha: Phys. Rev. A **40**, 4322 (1989); Phys. Rev. B **45**, 9541 (1992)
14. S.-L. Chang: *Multiple Diffraction of X-ray in Crystals* (Springer, Berlin Heidelberg New York 1984)
15. I.Ya. Dubovskaya, S.A. Stepanov, A.Ya. Silenko, A.P. Ulyanenkov: J. Phys.: Condens. Matter **5**, 7771 (1993)
16. I.D. Feranchuk: Kristallografia **24**, 289 (1979)
17. I.D. Feranchuk: Coherent phenomena in the processes of X-ray and gamma-radiation from relativistic charged particles in crystals. Habilitation Doctor Thesis, Belarussian State University, Minsk (1984)
18. I.D. Feranchuk, A.V. Ivashin, I.V. Polikarpov: J. Phys. D: Appl. Phys. **21**, 831 (1988)
19. R.B. Fiorito, D.W. Rule, M.A. Piestrup, X.K. Maruyama, R.M. Silzer, D.M. Skopik, A.V. Shchagin: Phys. Rev. E **51**, R2759 (1995)
20. Y. Hayakawa, M. Seto, Y. Maeda, T. Shirai, A. Noda: J. Phys. Soc. Japan **67**, 1044 (1998)
21. A. Kubankin, N. Nasonov, V. Sergienko, I. Vnukov: Nucl. Instrum. Methods B **201**, 97 (2003)
22. U. Pietsch, V. Holy, T. Baumbach: *High-Resolution X-ray Scattering. From Thin Films to Lateral Nanostructures* (Springer, New York 2004)
23. Z.G. Pinsker: *Dynamical Scattering of X-rays in Crystals* (Springer, Berlin Heidelberg New York 1981)
24. S.L. Smith, E.M. Purcell: Phys. Rev. **91**, 1069 (1953)

25. S.A. Stepanov, A.Ya. Silenko, A.P. Ulyanenkov, I.Ya. Dubovskaya: Nucl. Instrum. Methods B **117**, 44 (1996)
26. M.L. Ter-Mikaelian: *High Energy Electromagnetic Processes in Condensed Media* (in Russian: AN ArmSSR, Yerevan 1969) (in English: Wiley, New York 1972)

# 7

# Prospective Applications of PXR

## 7.1 PXR as a Tunable Source of Monochromatic X-rays

The previous chapters concern the fundamental properties of PXR, which have been experimentally observed and theoretically explained. The most recent efforts of scientists are aimed at finding effective applications of PXR, which profit from specific features of PXR.

The most prominent feature of PXR is a high radiation intensity in a narrow spectral interval, which is selected by experimental geometry or electron beam parameters. Therefore, the first application of PXR is seen as a monochromatic X-ray source. The most effective source of such kind is known to be a synchrotron radiation (SR), reflected from a monochromatizing crystal. The physical parameters of SR [48] are compared below with the analogous parameters of PXR. The angular distribution of photons emitted in the spectral interval $\Delta\omega$ per second and at the electron current $J$ is [48]

$$\left.\frac{\partial^2 N_{\mathrm{SR}}}{\partial\theta\partial\psi}\right|_{\psi=0} = \frac{3\alpha}{4\pi^2}\gamma^2\frac{\Delta\omega}{\omega}K(y)\frac{J}{e} \ . \tag{7.1}$$

The angle $\theta$ is defined in the orbital plane and is perpendicular to the electron velocity, whereas $\psi$ is normal to this plane; $K(y)$ is a universal function of the parameter $y = \omega/\omega_c$, which is related to the orbital radius $R$. This function describes the frequency distribution of radiation and reaches a maximum at

$$\omega = \omega_c = \frac{3\gamma^3 c}{2R} \ . \tag{7.2}$$

Usually, to compare different sources, relation (7.1) is transferred into practical units and considered as average brightness $\Upsilon$:

$$\begin{aligned}\left.\frac{\partial^2 N_{\mathrm{SR}}}{\partial\theta\partial\psi}\right|_{\psi=0} &= \Upsilon_{\mathrm{SR}}\ [\text{photons s}^{-1}[\text{mrad}]^{-2}(0.1[\%\ \text{bandwidth}])^{-1}] \\ &= 1.327\times 10^{13}\ E^2\ [\text{GeV}]\ J\ [\text{A}]\ K(y), \end{aligned} \tag{7.3}$$

and at frequency (7.2), the maximal energy of emitted photons is expressed via the amplitude $B$ of the magnetic the field:

$$\hbar\omega_c\ [\mathrm{keV}] = 0.665\ E^2\ [\mathrm{GeV}]\ B\ [\mathrm{T}]\,. \qquad (7.4)$$

According to Sect. 3.4, the optimal conditions for PXR are realized for dynamical diffraction of radiation in thick monocrystals of thickness $L > L_{\mathrm{abs}}$ and in the frequency and angular vicinity of Cherenkov's peak. Neglecting the effects of diffraction polarization, and for the frequency $\omega_{\mathrm{B}}$ corresponding to the reciprocal lattice vector $\boldsymbol{g}$ and the electron current $J_1$, (3.26)–(3.32) result in

$$\frac{\partial^3 N_{\mathrm{PXR}}}{\partial\theta\partial\psi\partial\omega} \approx \frac{\alpha}{\omega_{\mathrm{B}}\pi^2}(\theta^2+\psi^2)\left|\frac{\beta\chi_{\boldsymbol{g}}}{\delta_{\mathrm{B}}(\theta^2+\psi^2+\gamma^{-2}+\theta_{\mathrm{s}}^2-\chi_0-\delta_{\mathrm{B}})}\right|^2\frac{J_1}{e}\,,$$
$$\delta_{\mathrm{B}} = \frac{1}{2}\left[\mu+\sqrt{\mu^2+4\beta\chi_{\boldsymbol{g}}\chi_{-\boldsymbol{g}}}\right]\,,$$
$$\mu = \chi_0(\beta-1)-2\beta\left(\frac{\omega-\omega_{\mathrm{B}}}{\omega_{\mathrm{B}}}\frac{2\sin^2\theta_{\mathrm{B}}}{\cos 2\theta_{\mathrm{B}}}-\psi\tan 2\theta_{\mathrm{B}}\right)\,. \qquad (7.5)$$

The photons are concentrated in the cone with the centre of the reflection, determined by the angle $2\theta_{\mathrm{B}}$ to the electron velocity $\boldsymbol{v}$; the asymmetry parameter is $\beta > 0$ for the Laue case and $\beta < 0$ for the Bragg case; $\theta_{\mathrm{s}}^2$ is a mean-square angle of electron multiple scattering. For PXR, the angle $\psi$ defines the deviation of the photon wave vector from the reflection centre in the plane $(\boldsymbol{g},\,\boldsymbol{v})$ and $\theta$ is the deviation perpendicular to the $(\boldsymbol{g},\,\boldsymbol{v})$ direction; the polarizations $\chi_0, \chi_{\boldsymbol{g}}$ are calculated for the frequency $\omega = \omega_{\mathrm{B}}$.

As theory in Sect. 3.4 and experiment in Sect. 5.2 stated, the spectral width of the PXR peak is determined by the imaginary part of the X-ray polarizability. For the spectral interval $\Delta\omega > \omega_{\mathrm{B}}\chi_0'' \sim 10^{-6}$, (7.5) can be integrated over the frequency of detected photons and the analogue of (7.1) is obtained for PXR:

$$\left.\frac{\partial^2 N_{\mathrm{PXR}}}{\partial\theta\partial\psi}\right|_{\psi=0} = \frac{\alpha}{\pi}\frac{\beta^2\cos 2\theta_{\mathrm{B}}}{\sin^2\theta_{\mathrm{B}}} \times \frac{\theta^2|\chi_{\boldsymbol{g}}|^4}{|\chi_0''|\ |\theta^2+\gamma^{-2}+\theta_{\mathrm{s}}^2-\chi_0'|^4}\frac{J_1}{e}H[\Delta\omega-\omega_{\mathrm{B}}|\chi_0''|]\,, \qquad (7.6)$$

where $H[x]$ is a Heaviside function. The maximum of photon numbers is reached at an angle $\theta_0$ and the photons are concentrated in cone $\delta\theta$:

$$\theta_0^2 = \frac{1}{3}(\gamma^{-2}+\theta_{\mathrm{s}}^2-\chi_0')\,,$$
$$\delta\theta \approx \sqrt{\gamma^{-2}+\theta_{\mathrm{s}}^2-\chi_0'}\,. \qquad (7.7)$$

Then, the ratio of the average brightness for SR and PXR is

$$\xi = \frac{\Upsilon_{\mathrm{PXR}}}{\Upsilon_{\mathrm{SR}}} = \frac{9\pi}{64} \frac{\beta^2 \cos 2\theta_{\mathrm{B}}}{\sin^2 \theta_{\mathrm{B}}} \frac{|\chi_{\boldsymbol{g}}|^4}{|\chi_0''|\,|\gamma^{-2} + \theta_{\mathrm{s}}^2 - \chi_0'|^3} \frac{\omega_{\mathrm{B}}}{\gamma^2 \Delta\omega} \frac{J_1}{J} \,. \tag{7.8}$$

The analogue of (7.4) for PXR is the position of the central frequency of the PXR peak (1.20). In contrast to SR, this frequency is independent of energy; however, it depends on the interplane distance $d$, radiation angle and reflection order $n$. Contrary to SR, the photons are emitted not in a wide interval $K(y)$, but in a narrow interval $\delta\omega$, which depends on crystal parameters only:

$$\begin{aligned} \hbar\omega_{\mathrm{B}}\ [\mathrm{keV}] &= 1.974 \frac{\pi}{d[\text{Å}] \sin\theta_{\mathrm{B}}} n \,, \\ \delta\omega &\approx |\chi_0''| \frac{\cos 2\theta_{\mathrm{B}}}{2\sin^2\theta_{\mathrm{B}}} - \delta\psi \cot\theta_{\mathrm{B}} \,. \end{aligned} \tag{7.9}$$

Thus, the principal advantage of PXR from relativistic particles is, according to (7.7) and (7.9), a narrow and highly monochromatic photon beam, which is formed directly during emission and does not require any further monochromatizing, as in the SR case [29]. Moreover, the PXR features do not depend on the particle energy if the following conditions are fulfilled:

$$\gamma^{-2} < |\chi_0'|, \qquad E > E_{\mathrm{opt}} = mc^2 \sqrt{|\chi_0'|} \sim 50\text{–}100\,\mathrm{MeV} \,. \tag{7.10}$$

For this particle energy, the photons with $\hbar\omega \sim 1\text{–}20\,\mathrm{keV}$ can be produced, whereas for SR the electron energy must be $E \sim 2\text{–}5$ GeV to yield the photons of the same frequency.

The quantum output $Q(E)$ is often used to quantify the absolute value of the average brightness of a PXR source in the units of (7.3). The value $Q$ is defined as the number of photons emitted by one electron of energy $E$ at a solid angle $\Delta\Omega_{\mathrm{D}} = 1\ \mathrm{mrad}^2$ within the spectral PXR peak. Under conditions (7.5), this value is

$$\begin{aligned} Q(E)\ [\text{photons e}^{-1}\ \text{mrad}^{-2}] &\approx \frac{\alpha}{\pi} \frac{\beta^2 \cos 2\theta_{\mathrm{B}}}{\sin^2\theta_{\mathrm{B}}} \\ &\times \int \mathrm{d}\Omega_{\mathrm{D}} \frac{(\theta^2 + \psi^2)|\chi_{\boldsymbol{g}}|^4}{|\chi_0''|\,|\theta^2 + \gamma^{-2} + \theta_{\mathrm{s}}^2 - \chi_0'|^4} \,. \end{aligned} \tag{7.11}$$

Then, the PXR average brightness in the above-mentioned units is

$$\begin{aligned} \Upsilon_{\mathrm{PXR}}\ &[\text{photons s}^{-1}\ \text{mrad}^{-2} (0.1\%\ \text{bandwidth})^{-1}] \\ &\approx 0.625\, Q(E) \times 10^{22}\ J_1\ [\mathrm{A}] \,. \end{aligned} \tag{7.12}$$

The magnitude of $Q(E)$ under condition (7.10) is weakly dependent on the electron energy; however, it is strongly influenced by a crystal type and experimental geometry. The estimates made in Sects. 2.2 and 3.4 (see also Fig. 7.1) give the limits

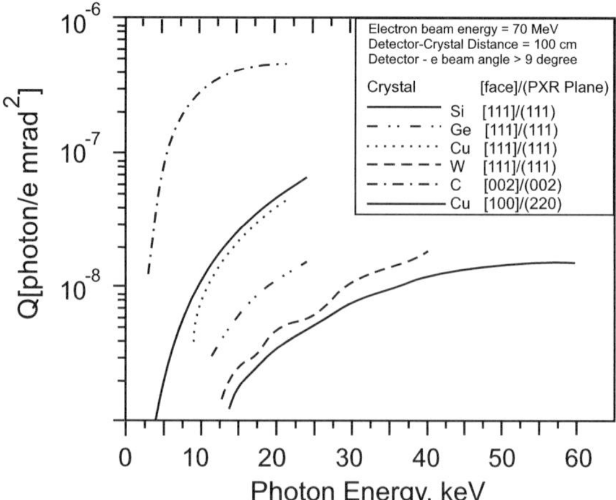


**Fig. 7.1.** Calculated quantum output for 500-μm-thick crystal targets for several reflection planes and crystal face planes [17]. The upper photon energy limit in each case was limited by a requirement of an angle of more than 9° between the detector and the electron beam axis

$$10^{-8} \leq Q(E) \leq 10^{-4} \,. \tag{7.13}$$

The practical significance of the PXR source becomes apparent either when its average brightness is comparable with SR brightness, or when the compactness and low power inputs are important. The analysis of (7.8) indicates several ways for optimization of the PXR source: (i) the choice of the crystal and the reflection to maximize the quantum output $Q(E)$; (ii) the choice of electron source and radiation geometry to reach high current $J_1$ for energy $E \sim E_{\text{opt}}$; (iii) selection of the applications requiring narrow $\Delta\omega$. Up to now, a large volume of theoretical and experimental work has been done in the mentioned directions.

The diverse materials have been investigated at Rensselaer Polytechnic Institute (USA) for construction of a novel tunable X-ray source using the 100 MeV linear accelerator (see [17,39,40] and citations therein). The objective of their investigation is the optimized production of PXR for future applications in medical imaging, material characterization and detection of explosives and nuclear materials. Up to now, the main interest is connected with optimization of the crystal characteristics: lattice parameters; bulk thickness; electric, thermal and absorptive properties; the growth and polishing techniques. Numerous theoretical simulations and experiments have been done to optimize the mentioned characteristics; some of the results are presented in Fig. 7.1.

A variety of well-documented single crystals to include Si, Ge, Cu, pyrolytic graphite and W was considered. PXR intensities were calculated for the (111), (220) and (002) planes at Bragg angles from 5° up to 90°. The differential intensity at the peak of one of the angular distribution cones was integrated across the 1 $mm^2$ detector surface placed at a distance 1 m from the crystal. For each crystal, the peak intensity appears at a different Bragg angle and different X-ray energies. Pyrolytic graphite shows to be most promising for photon energies less than 25 keV, while Cu or W is best at higher energies. However, the mosaic spread of the graphite might degrade these results and this issue is now under experimental investigations by authors of [162–175].

The recent experiments ([2,30] and citations therein) at synchrotron Sirius (Tomsk, Russia) are principal in the case of the increase of the average PXR brightness. Most of PXR experiments (Chaps. 5 and 6) have been conducted at linear accelerators, where the average current $J_1 \leq 10^{-6}$ A, which is much less than that at synchrotron $J_1 \sim 1$ A. The experiment [2] has been carried out for investigation of the PXR emission during the multiple passes of electrons through periodic and crystalline targets mounted inside a synchrotron, and the similar experiments [35] have been carried out at SAL (Canada).

The arrangement of the experiment [2] is shown in Fig. 7.2. The electrons strike the internal radiator $T$. The current of accelerated electrons $N_e$ with the initial energy $E = 800\,\mathrm{MeV}$ was measured with an indicative device $I$. Bremsstrahlung $B_\mathrm{s}$, generated in the target, was detected by a Gauss-quantometer $Q$. In the case of the crystal target, the PXR emitted at an angle of $\theta_\mathrm{D} = 18°12'$ with respect to the electron beam was observed by a NaI(Tl) detector collimated by a vertical slit, which provided an angular resolution of about 1 mrad. The scraper $S$ on the opposite side of the synchrotron ring intersected electrons which had lost energy after passing through the internal radiator. The quantometer determined the radiation losses, $W_e$, of the electrons in the target as a function of the radial position $R$ of the target. From the experimental data for the energy losses, the number of passes $k$ of the electrons through the target was estimated in the range 10–46 for various targets.

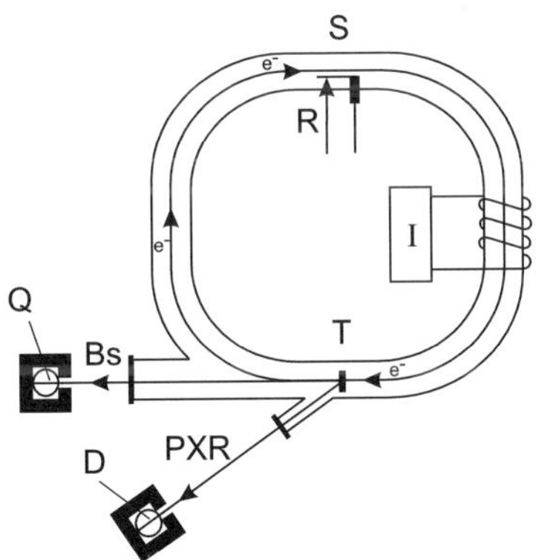


**Fig. 7.2.** Experimental arrangement from [2]: target $T$, Gauss quantometer $Q$, NaI(Tl)-spectrometer $D$, scraper $S$, indicative device $I$

These data show that multiple passes through targets may increase the brightness of the PXR source mounted inside cyclic accelerators or storage rings. For the investigation of the influence of multiple passes of the electron beam on the angular and spectral characteristics of PXR, a 48-μm-thick (220) Si crystal mounted on a goniometer head was used [2]. The experimental geometry corresponded to the first PXR spectral-peak position $\omega_B \approx 19$ keV. The measured angular and frequency spectra of the collimated PXR emitted from the crystal aligned under the Bragg condition $\theta_B = 0.5\theta_D$ are shown in Fig. 7.3 for a single pass, $k = 1$ (points) and for multiple pass, $k = 20$ (solid curves). The spectra are normalized at their peaks in order to compare their spectral distributions, but the absolute yield was increased by a factor of about 20.

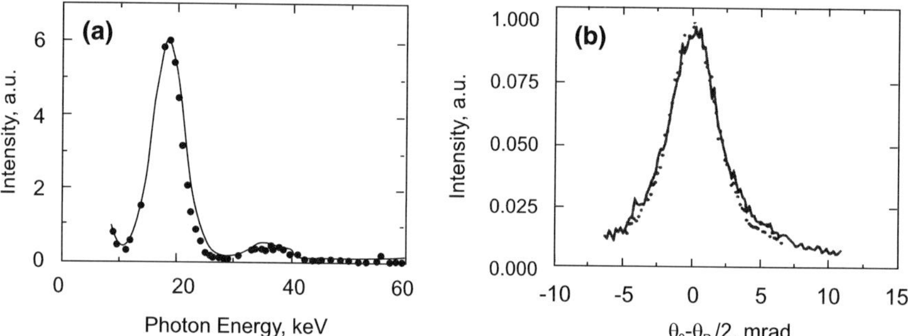


**Fig. 7.3.** Peak-normalized spectra [2] of collimated PXR generated from a single pass, $k = 1$ (points) and multiple passes, $k = 20$ (solid line) of 800 MeV electrons through a 48-$\mu$m-thick (220) Si crystal: (**a**) frequency distribution; (**b**) angular spectra

The experiment [30] also demonstrated the brightness gain for PXR from artificial periodic structures. The scheme of the experimental set-up [30] is shown in Fig. 7.4. The synchrotron has a 20 ms pulse duration with a 4 Hz repetition rate, and the divergence of the electron beam was about 0.2 mrad; the radiator was mounted on a goniometer in the Bragg geometry.

The X-ray mirror, used as a periodic structure, consisted of 300 periods of W and $B_4C$ with spacing $d = 12.36$ Å, which have been grown on a 380-μm Si substrate. The W and $B_4C$ layers were of thicknesses $d_1 \approx 5$ Å and $d_2 \approx 7$ Å, respectively. The X-rays generated in the mirror or in its Si substrate were detected by a CdTe semiconductor detector placed at an angle $\theta_D = 66.3$ mrad with respect to the electron beam velocity at a distance of 443 cm. The detector aperture was 4 mm$^2$ and the relative energy resolution reached 10% at line 8.1 keV. The bremsstrahlung $\gamma$-rays generated in the mirror were detected with a gamma quantometer and used to determine the current through the mirror and, ultimately, the mirror efficiency.

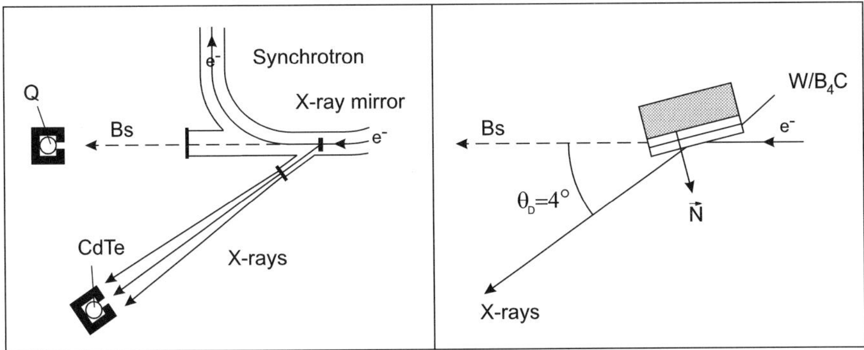


**Fig. 7.4.** The experimental apparatus from [30]: $\gamma$-quantometer $Q$, detector $D$, radiator $T$ mounted on the goniometer head

Both the spectra of the X-rays generated in the mirror and in the Si crystalline substrate were measured. The results of [30] in Fig. 7.5 demonstrate the source brightness gain for multilayers, compared to a monocrystal, despite the fact that the thickness of the multilayers (0.37 µm) is less than that of the crystal (100 µm). The quantum output $Q(E)$ for [30] and the average brightness are

$$\begin{aligned} Q(E) &\approx 2.2 \times 10^{-7} \text{ [photons e}^{-1}\text{ mrad}^{-2}] , \\ \Upsilon \text{ [photons s}^{-1}\text{ mrad}^{-2}&(0.1\%\text{ bandwidth})^{-1}] \\ &\approx 1.375 \ \times 10^{15} \ J_1 \text{ [A]} , \end{aligned} \tag{7.14}$$

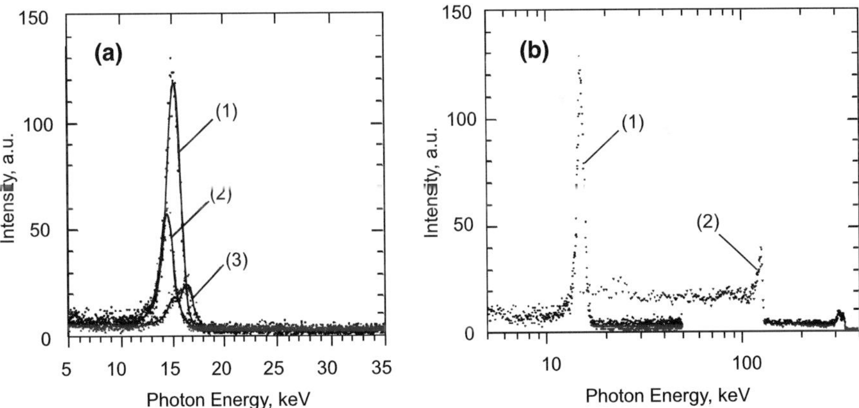


**Fig. 7.5.** The spectra measured in [30]: (**a**) the X-rays generated in the multilayers at mirror angles $\theta_0 = 31.1, 33.2, 36.1$ mrad (curves 1, 2 and 3, respectively), maximal peak (1) corresponds to the Bragg angle of the reflection $\theta_0 = \theta_\mathrm{B} = \theta_\mathrm{D}/2$; (**b**) spectra, generated in the mirror (1) and in the Si substrate

which are satisfactory for some medical applications [37].

To find optimal conditions for the PXR brightness gain from artificial periodic structures [41,42,45], we consider here the basic principles of emission by a relativistic particle in a one-dimensional periodic structure. The total intensity is an interference of transition radiation [34] from periodic interfaces between different media. When the polarizability of the multilayers $\chi_j \leq 10^{-5}$, this radiation is shown [11, 23] to be equivalent to PXR from a one-dimensional crystal with periodic polarizability:

$$\chi(\boldsymbol{r},\omega) = \sum_{l=-\infty}^{\infty} \chi_l(\omega) e^{i\boldsymbol{g}_l \boldsymbol{r}}, \qquad \chi_0(\omega) = \frac{\chi^{(W)} d_1 + \chi^{(B_4C)} d_2}{d},$$
$$\chi_l(\omega) = \frac{i(\chi^{(B_4C)} - \chi^{(W)})}{2\pi l}(1 - e^{-i g_l d_1}),$$
$$\chi_{-l}(\omega) = -\frac{i(\chi^{(B_4C)} - \chi^{(W)})}{2\pi l}(1 - e^{i g_l d_1}),$$
$$\boldsymbol{g}_l = g_l\, \boldsymbol{N}, \qquad g_l = \frac{2\pi}{d} l, \qquad l = 0, \pm 1, \pm 2, \dots, \tag{7.15}$$

where $\boldsymbol{N}$ is a normal vector to the surface of multilayers (Fig. 7.4), $\chi^{(W)}(\omega)$, $\chi^{(B_4C)}(\omega)$ are the X-ray polarizabilities of tungsten and $B_4C$, respectively. Using these definitions, the PXR from the multilayers can be calculated using formulas for PXR in the crystal. For instance, the PXR frequencies $\omega_B^{(1)}$ for the first multilayers harmonics and the $\omega_{113}$ PXR reflection for Si substrate at $\theta_B = 33.15$ mrad are

$$\hbar\omega_B^{(1)} = \frac{\pi}{d \sin\theta_B} \approx 15.4 \text{ keV},$$
$$\hbar\omega_{113} = \frac{\pi\sqrt{11}}{d_{Si} \sin\theta_B} \approx 118 \text{ keV}, \tag{7.16}$$

which fit the experimental results in Fig. 7.5. The theoretical estimate for the quantum output $Q(E)$ in [30] follows from (7.11) at $\Delta\Omega_D = 1$ mrad$^2$:

$$Q(E)\ [\text{photons } e^{-1} \text{ mrad}^{-2}] \approx \frac{\alpha}{\pi \sin^2\theta_B} |\chi_{\boldsymbol{g}}(\omega_B)| \omega_B \frac{2L}{c}\, 10^{-6}, \tag{7.17}$$

where the radiator thickness $L < L_{abs}$, as was in [30]. The effective polarizability of the multilayers is calculated on the basis of the layers' polarizabilities [19, 43]:

$$|\chi_{\boldsymbol{g}}^{ML}(\omega_B^{(1)})| \approx 2.9 \times 10^{-6},$$

and

$$Q(E) \approx 3.3 \times 10^{-7}\ [\text{photons } e^{-1} \text{ mrad}^{-2}],$$

which agrees with the results of [30].

According to (7.17), the essential gain of $Q(E)$ is due to the small scattering angle $\theta_B$ of photons. Thus, both the choice of radiator crystal and that of photon emission angle are important for optimization of the PXR the source.

The energy $E$ of the electron beam influences the effectiveness of PXR as well. The dependence $Q(E)$ is conditioned by multiple factors (see Sect. 5.3); however, the qualitative estimate from [25] gives a satisfactory accuracy for average brightness:

$$\Upsilon_{\mathrm{PXR}}(E) \approx \frac{E^2}{E^2 + E_{\mathrm{opt}}^2} \, Q(E_{\mathrm{opt}}) \times 10^{22} \, J_1(E) \, [\mathrm{A}] \,, \tag{7.18}$$

where $E_{\mathrm{opt}}$ is defined in (7.10).

The decrease of the quantum output at low electron energy can be compensated in average brightness by the increase of the current $J_1(E)$, as for example, in compact betatron yielding $E \sim 30\,\mathrm{MeV}$ electrons and designed for mammography [37]. The PXR from nonrelativistic electrons ($E \sim 100\,\mathrm{keV}$) is suggested in [25] for the increase of the radiation brightness (see Chap. 4), which is important at laboratory conditions [28].

## 7.2 Anomalous Scattering Method for Crystallography

The PXR spectra include information about the structures where this radiation is generated, which promises one more application of PXR. The analysis of these spectra could complete essentially the standard methods based on the conventional diffraction of external X-rays in the investigated crystals [21,24] or periodical multilayered structures [22,44]. In this section, the possibility of the PXR application for the direct measurement of the phases of structure amplitudes is considered. This problem is very essential for unambiguous calculation of the electron density in the X-ray structure analysis.

Development of various physical methods for the direct measurements of the phases of structure amplitudes is of great interest for X-ray diffractometry [46,47]. These methods are especially claimed when the structure of the organic crystals with a complicated multiple-atom elementary unit cell has to be determined. Several simple and effective methods for solution of the phase problem were suggested long time ago for special cases (see [46] and citation therein). The first is the method of the isomorphic replacement when a heavy atom is introduced into the investigated structure. The second approach is the anomalous scattering method (ASM) when the intensities of the reflected waves are measured for two different X-ray wavelengths close to the absorption edge of one of the atoms in the crystallographic unit cell. Both methods permit us to obtain the phase information even for thin or mosaic crystals when the kinematic theory of diffraction is applicable. For the practical X-ray structure analysis, the phases can be calculated through the solutions of the simple algebraic equations is also essential.

Several methods for the direct measurement of the phases have been suggested on the basis of the interference phenomena which are related to the various multiple-beam diffraction set-ups. As a rule, the interference between the waves diffracted by different planes is essential for thick and perfect crystals when the effects of the dynamical X-ray diffraction are strongly pronounced. However, the realization of such methods for macromolecular organic crystals is very problematic because most of them are thin and mosaic. Besides, the evaluation of the phases in the case of multiple-beam diffraction even for the ideal crystals requires complicated and precise measurements.

The further development of simple and universal realizations of ASM is still attractive due to its kinematical domain. There are two known obstacles for the universal application of ASM: (i) an X-ray source of monochromatic radiation with smoothly varied wavelength has to be used and (ii) absorption edges in most organic crystals correspond to soft X-rays. For example, the most heavy sulfur atom in amino acids has the $K_\alpha$-line corresponding to the photon energy $\hbar\omega_K \simeq 2.45\,\text{keV}$, or the wavelength $\lambda_K \simeq 5.019$ Å. Both problems are essential for ASM application under the laboratory conditions because of the absence of tunable soft X-ray sources. They can be overcome with the use of the synchrotron radiation; however, there are also some difficulties in this case, i.e. the standard crystal monochromators are not efficient for soft X-rays.

The ASM realization discussed in the present section is based on the analysis of the spectra of the parametric X-ray radiation generated by the electrons passing through the investigated crystal. This approach could be realized under the laboratory conditions with PXR either from relativistic electrons ($E \sim 30\text{–}50\,\text{MeV}$), when the crystal is placed inside the compact betatron [37], or from nonrelativistic electrons ($E \sim 100$ keV).

As follows from formula (2.47), the radiation frequency in the PXR reflection can be tuned by simple rotation of the crystal with respect to the electron beam. This PXR feature is analysed here in order to justify the possibility of the direct measurement of the phase of the structure amplitude $F(\boldsymbol{g})$. We choose the incidence angle of the electron beam in such a way that the radiation frequency of the PXR reflection from the investigated plane is equal to the frequency $\omega_K$ corresponding to the $K$-edge of the absorption band of one of the substances in the considered crystal:

$$\sin\theta_{\mathrm{B}}^{K} = \frac{gc^2}{2v\omega_K} = \frac{c}{4\pi v} g\lambda_K \,, \tag{7.19}$$

where $\lambda_K$ is the wavelength of the $K_\alpha$-line for this substance.

An analogous approach has been discussed in [21] for the case of the dynamical diffraction and hard X-rays corresponding to the characteristic lines of heavy atoms. However, these conditions are not fulfilled for the crystals of the organic compounds, which are of the most interest in the current investigation. Therefore, we estimate the typical parameters of the interaction between X-rays and organic crystals, i.e. the polarizability $\chi_0$, the extinction length $L_{\text{ext}}$ and the absorption length $L_{\text{abs}}$. For numerical calculations, the organic compound with the chemical formula

$$C_{13}H_{14}N_2OS \tag{7.20}$$

is chosen, whose structure has been recently resolved [27] . The crystal possesses a monoclinic symmetry described by the space group $P2_1/c$ and includes four molecules in the elementary unit cell. The parameters of the cell $(a, b, c)$, the volume $\Omega$ and the angle $\beta$ are

$$a = 10.202\,\text{Å}, \qquad b = 7.203\,\text{Å}, \qquad c = 17.618\,\text{Å}\,;$$
$$\Omega = 1270\,\text{Å}^3; \quad \beta = 101.09^\circ\,. \tag{7.21}$$

In accordance with formula (2.47), most of the PXR reflections for this crystal correspond to large wavelengths $\lambda \leq 40\,\text{Å}$. The wavelength of the $K$-edge of the sulfur atom belongs to this range and is considered as the basic for the realization of the ASM scheme. Another reason for the selection of sulfur is the presence of this element in most of organic compounds; thus isomorphic replacement is not required. The atom of sulfur has also a large nucleus charge and, therefore, its coordinates are well defined in the X-ray structure analysis [38].

The total number of electrons in the elementary unit cell of the considered crystal is $Z_{\text{tot}} = 4 \times 138 = 552$. The X-ray polarizability $\chi_0(\lambda_1)$ at the wavelength $\lambda_1 = \lambda_K/1.1 \simeq 4.66\,\text{Å}$ near $\lambda_K \simeq 5.019\,\text{Å}$ corresponding to the $K$-edge of the absorption band of the atom S is

$$\chi_0 = -\frac{e^2 Z_{\text{tot}} \lambda_1^2}{4\pi^2 mc^2 \Omega}\,, \tag{7.22}$$

and the numerical values are

$$\chi_0 \simeq -7.98 \times 10^{-5} + \mathrm{i}\, 3.47 \times 10^{-6}; \qquad L_{\text{ext}} = \frac{\lambda_1}{\pi|\chi_0|} \simeq 1.86\ \mu\text{m};$$
$$L_{\text{abs}} \simeq 21.4\ \mu\text{m}\,, \tag{7.23}$$

The average effective charge for a single scattering atom in the organic crystals is small enough (for the considered case $Z_{\text{eff}} \simeq 4$), therefore, electron multiple scattering in an organic matter is essentially less than that in inorganic crystals. The organic crystals are usually grown as thin films or mosaic crystals consisting of small blocks of thickness $L \sim L_{\text{ext}}$ . Thus, the angular and frequency distributions of the PXR intensity emitted by one electron into the reflection with the vector $\omega_{\text{B}} \boldsymbol{v}/c^2 + \boldsymbol{g}$ can be calculated in the framework of kinematic approximation. The frequency of X-rays in the reflection unambiguously depends on the exit angle of the photons because of condition (2.54), and the angular distribution within the reflection is defined by the following parameter:

$$\theta_{\rm ph} = \sqrt{\gamma^{-2} + |\chi_0|} \simeq 10^{-2} \, . \tag{7.24}$$

The main contribution to the reflection intensity is defined by the first non-forbidden harmonic. The spectral width of this harmonic depends on the parameter $\theta_{\rm ph}$:

$$\left(\frac{\Delta\omega}{\omega}\right)_{\rm PXR} \simeq \theta_{\rm ph} \, . \tag{7.25}$$

To perform an integration over the angles in formula (2.54), the dependence of polarizability on frequency and angle due to equation (7.19) should be taken into account. This dependence has an obvious form (A.1):

$$\chi_{\boldsymbol{g}}(\omega) = -\frac{4\pi r_e}{\omega^2 \Omega} F(\boldsymbol{g}) \, ;$$
$$F(\boldsymbol{g}, \omega) = \sum_j [f_{0j}(\boldsymbol{g}) + f'_j(\omega) + \mathrm{i}\, f''_j(\omega)] \mathrm{e}^{-W_j(\boldsymbol{g})} \mathrm{e}^{\mathrm{i}\boldsymbol{g}\boldsymbol{R}_j} \, . \tag{7.26}$$

Within the range of the PXR harmonic spectral width, the most essential dependence on the frequency in formula (1.13) is defined by the functions $f', f''$. This dependence becomes universal and localized near the frequency $\omega \simeq \omega_K$ of the absorption edge [46], if the anomalous corrections are normalized to $f_0$. Figure 7.6 demonstrates the relative width of the absorption edge, which is essentially more than the characteristic width of the PXR reflection $\Delta\omega/\omega_K \simeq 1 \gg \theta_{\rm ph}$. Therefore, the polarizability $\chi_{\boldsymbol{g}}(\omega)$ is assumed to be constant with the argument $\omega = \omega_{\rm B}^{(n)}$ for calculation of the integral in (1.7). The value $f''(\omega)$ is not well determined at the angle $\theta_H$, which corresponds to the condition $\omega_{\rm B}^{(n)} = \omega_K$. However, the applications considered below operate in the frequency range

$$1.1 < \left| \frac{\omega_H^{(n)} - \omega_K}{\omega_K} \right| < 1.5 \, , \tag{7.27}$$

where the anomalous corrections give an important contribution to the structure amplitudes, changing smoothly within the limits of the PXR reflections.

After integration over the angles in formula (2.54), the intensity of the photons $I_n$, emitted into the single PXR harmonic by the electron current $J$ passing through the investigated crystal, is calculated straightforwardly:

$$I_n = A(E, \omega_{\rm B}^{(n)}) |F(\boldsymbol{g}_n, \omega_{\rm B}^{(n)})|^2 \, ;$$

$$A(E, \omega_{\rm B}^{(n)}) = \frac{e^2 2\pi^2 r_e^2}{\left(\omega_{\rm B}^{(n)}\right)^3 \Omega^2 \hbar c^4} \frac{(1 + \cos^2\theta_{\rm B}) L_{\rm abs}}{\sin^2\theta_{\rm B}}$$
$$\times \left(1 - \exp\left(-\frac{L}{L_{\rm abs}}\right)\right) \left[\ln\left(\frac{\theta_{\rm D}^2 + \theta_{\rm ph}^2}{\theta_{\rm ph}^2}\right) - \frac{\theta_{\rm D}^2}{\theta_{\rm D}^2 + \theta_{\rm ph}^2}\right] \frac{J}{e} \, . \tag{7.28}$$

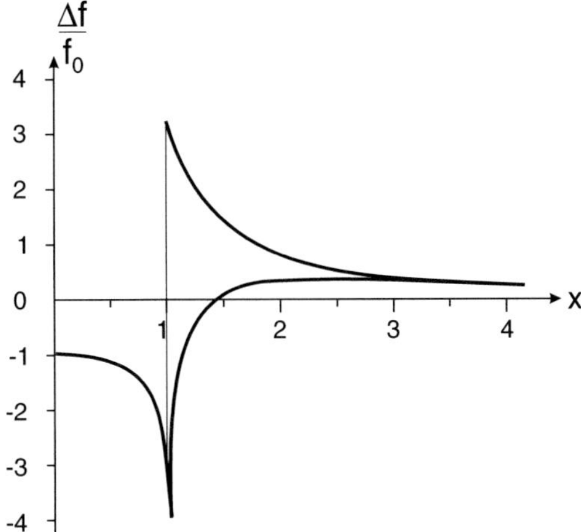


**Fig. 7.6.** Imaginary (*upper curve*) and real (*lower curve*) anomalous corrections in the range of the absorption edge

Here the parameter $\theta_{\mathrm{D}} \sim \theta_{\mathrm{ph}}$ defines the angular width of the detector and the harmonic spectral width is in accordance with formula (7.25). The spectral resolution of the detector is supposed to be sufficient to measure separately every PXR harmonic. The value $A(E, \omega_H^{(n)})$ in (1.15) describes the kinematics of the radiation process and does not depend on the structure amplitude $F$.

Using fundamental constants, formula (7.28) permits us to estimate the rate of the photon counts in experiments:

$$I_n \left[\frac{\text{photon}}{\text{s}}\right] \simeq 5.6 \times 10^7 \frac{1}{(\omega_{\mathrm{B}}^{(n)})^3 \Omega^2} K |F(\boldsymbol{g}_n, \omega_{\mathrm{B}}^{(n)})|^2 J\,[\mu\mathrm{A}]\ ;$$

$$K = \frac{(1+\cos^2\theta_{\mathrm{B}})L_{\mathrm{abs}}}{\sin^2\theta_{\mathrm{B}}} \left[1 - \mathrm{e}^{-L/L_{\mathrm{abs}}}\right] \left[\ln\left(\frac{\theta_{\mathrm{D}}^2 + \theta_{\mathrm{ph}}^2}{\theta_{\mathrm{ph}}^2}\right) - \frac{\theta_{\mathrm{D}}^2}{\theta_{\mathrm{D}}^2 + \theta_{\mathrm{ph}}^2}\right]. \quad (7.29)$$

The following units are used: keV for radiation frequency, Å$^3$ for the volume of the crystallographic unit cell, $\mu$m for the absorption length and the crystal thickness and $K$ is approximately equal to unity. For the crystal (7.20) of thickness $L = 10\,\mu\mathrm{m}$ and X-ray radiation close to the $K_\alpha$-line of the atom S for reflection with $F(\boldsymbol{g}) \sim 20$, formula (7.29) gives an estimate

$$I_n \left[\frac{\text{photon}}{\text{s}}\right] \simeq 1.0 \times 10^4 J\,[\mu\mathrm{A}]\ . \quad (7.30)$$

The success of experimental realization of the considered method depends on the ratio of the X-ray intensity in the PXR reflection (signal) to the intensity of X-rays emitted by the electrons due to another processes, which are not strictly connected to the structure amplitudes (noise). The bremsstrahlung is the process contributing essentially to radiation background to the continuous spectrum due to the interaction between the electrons and medium. In the case of relativistic electrons, the total flux of the bremsstrahlung is concentrated in the narrow cone with the angle of divergence $\sim \theta_{\mathrm{ph}}$ along the particle velocity. Therefore, the photons emitted into PXR reflections at the large angles with respect to the electron velocity are registered with low background, as confirmed in several experiments.

The only competing process for PXR in the range of the anomalous scattering is a characteristic radiation (ChR), which is caused by excitation of the crystal atoms by electron beam. This radiation is represented by the set of narrow spectral lines and has almost isotropic angular distribution. In most of the cases, the relative spectral width of the ChR lines is essentially less than the width of the PXR harmonics defined by formula (7.25):

$$\left(\frac{\Delta\omega}{\omega}\right)_{\mathrm{ChR}} \ll \left(\frac{\Delta\omega}{\omega}\right)_{\mathrm{PXR}} . \tag{7.31}$$

If the photons are detected by an X-ray spectrometer with the energy resolution of the same order as the spectral width of the PXR reflection, the intensities of PXR and ChR can be independently measured in the frequency range (7.27). To compare the radiation intensities for both mechanisms, the experimental data for the photo-ionization $\sigma_{\mathrm{ph}}$ and for the fluorescence yield $Y_f(Z)$ [48] are used for calculation of the probability of the excitation of the atom S by the electron beam ($Z$ is the atom nucleus charge). Expression (7.19) is used for calculation of the spectral density of pseudophotons (7.30) corresponding to one electron. Then the intensity of ChR quanta generated from the crystal with thickness $L$ and the density of the sulfur atoms $n_{\mathrm{S}}$ by the electron current $J$ and registered by the detector with the angular resolution $\theta_{\mathrm{D}}$ can be calculated as follows:

$$I_{\mathrm{ChR}} \simeq \int_{\omega_{\min}}^{\infty} \mathrm{d}\omega n(\omega)\sigma_{\mathrm{ph}}(\omega)Y_f(Z)n_{\mathrm{S}}L\frac{\theta_{\mathrm{D}}^2}{2}\frac{J}{e} . \tag{7.32}$$

Experimental data for the value $\sigma_{\mathrm{ph}}$ of the atom S can be interpolated by the analytical formula:

$$\sigma_{\mathrm{ph}}(\omega) \simeq \sigma_0 \left(\frac{\omega_{\min}}{\omega}\right)^2 ; \quad \sigma_0 \simeq 0.7 \times 10^{-22}\ \mathrm{m}^2; \quad \hbar\omega_{\min} \simeq 150\ \mathrm{eV} ,$$

and the value $Y_f(16) \simeq 0.1$ [48]. We choose the same values for other physical parameters as in expression (7.30):

$$n_{\mathrm{S}} = \frac{1}{\Omega}; \quad \Omega = 1270\,\mathring{\mathrm{A}}^3; \quad L = 10\,\mu\mathrm{m}; \ \theta_{\mathrm{D}} = 10^{-2} .$$

Finally,

$$I_{\mathrm{ChR}} \left[\frac{\mathrm{photon}}{\mathrm{s}}\right] \simeq 2.1 \times 10^4 J\ [\mu\mathrm{A}] . \tag{7.33}$$

Thus, the intensities of ChR and PXR have the same order of magnitude for the considered experimental conditions. The high intensity of PXR is essential for the following experimental aspect of the problem. The anomalous scattering corrections for the atom S have the values $3 < |\Delta f'| \sim \Delta f'' < 5$ for the frequency range (7.27), which corresponds to several percents from the full structure amplitude of typical organic crystals. Therefore, the use of ASM requires accurate measurement of the intensity of PXR reflections, which also takes place for the standard X-ray structure analysis [38]. The accuracy of the intensity measurements is mainly defined by the fluctuations of the counting rate, which produces a restriction for the minimal observation time $t_{\min}$. Particularly, in order to measure the structure amplitude with the relative accuracy $\delta_F$, the value $t_{\min}$ should be of order

$$t_{\min} \simeq \frac{1}{I_n \delta_F^2} , \tag{7.34}$$

where $I_n$ is the X-ray intensity.

Now we consider how this information can be used for the direct measurement of the phases of structure amplitudes. We provide the simulation for the organic crystal described by (7.20) and (7.21) in order to find the experimental conditions for realization of ASM on the basis of PXR.

The structure of the organic crystal (7.20) has already been solved [27], and the real values of the structure amplitude modulus $|F_0(\boldsymbol{g})|$ and its phase $\varphi_0(\boldsymbol{g})$ can therefore be calculated for the reciprocal lattice vector $\boldsymbol{g} = (h, k, l) = (1, 1, 3)$. When this amplitude is calculated for the wavelength $\lambda$, which is essentially different from the $K$-edge $\lambda_K$ of the absorption edge of any atom in the crystal unit cell, the anomalous scattering corrections are small and the structure amplitudes do not depend on the wavelength but are defined by the electron density distribution only. The following values are calculated for the above-mentioned reflection and $\lambda = 1.54$ Å [19, 43]:

$$\begin{aligned} F_0(1,1,3) &= F_0' + \mathrm{i}\, \mathrm{F}_0'' = 11.79 - \mathrm{i}\; 0.170 \; ; \\ |F_0| = 11.79; \quad \varphi_0 &= \arctan \frac{F_0''}{F_0'} = -0.014 \; . \end{aligned} \tag{7.35}$$

When ASM is used, the modulus of the structure amplitude $|F_0(\boldsymbol{g})|$ is supposed to be known because it can be measured from either corresponding diffraction intensity or PXR intensity in the same reflection but far from the absorption edge. The structure factor $|S(\boldsymbol{g})|$ and its phase $\psi(\boldsymbol{g})$ for the sulfur atom in the crystallographic unit cell are defined in the following way:

$$F_{\mathrm{S}}(\boldsymbol{g}) = f_{\mathrm{S}}(\boldsymbol{g}) S(\boldsymbol{g}), \quad S(\boldsymbol{g}) = \mathrm{e}^{-W_{\mathrm{S}}(\boldsymbol{g})} \sum_{\mathrm{s}} \mathrm{e}^{\mathrm{i}\boldsymbol{g}\boldsymbol{R}_{\mathrm{s}}} \tag{7.36}$$

where $f_{\mathrm{S}}(\boldsymbol{g})$ and $\mathrm{e}^{-W_{\mathrm{S}}(\boldsymbol{g})}$ are the atomic scattering factor and the Debye–Waller factor of the sulfur atom, respectively, and the structure factor $S(\boldsymbol{H})$

is defined by the coordinates $\boldsymbol{R}_s$ of these atoms in the elementary cell. For the considered reflection $\boldsymbol{g} = (1, 1, 3)$ and wavelength 1.54 Å these values are equal to

$$f_S(\boldsymbol{g}) = 1.718 - i0.069; \quad e^{-W_S(\boldsymbol{g})} = 0.94 ;$$
$$|S(\boldsymbol{g})| = 0.133 \quad \psi(\boldsymbol{g}) \simeq 0 . \tag{7.37}$$

Usually, the modulus and phase of the structure factor $F_S(\boldsymbol{H})$ can be considered as known value because the positions of the most heavy atoms in the unit cell are defined accurately during the primary analysis of the Patterson function [38]. This assumption on the values $|S(\boldsymbol{g})|, \psi(\boldsymbol{g})$ actually means that unknown phases are calculated relatively to the phases corresponding to the distribution of the anomalously scattering atoms in the unit cell. Therefore, the phase $\psi(\boldsymbol{g})$ can be substituted by the zero value.

Now the structure amplitudes may be calculated taking into account the anomalous scattering corrections for the sulfur atom and the X-ray frequencies $\omega_1 = 1.1\omega_K$; $\omega_2 = 1.5\omega_K$. The results are as follows:

$$F(\boldsymbol{g}, \omega_{1,2}) = |F_0(\boldsymbol{g})|(\cos\varphi_0 + \mathrm{i}\sin\varphi_0) + \Delta \mathrm{F}'_{1,2} + \mathrm{i}\Delta \mathrm{F}''_{1,2} ;$$
$$\Delta F_{1,2} = \Delta f_S(\omega_{1,2}) S(\boldsymbol{g}) ;$$
$$\Delta F''_1 = -0.696; \quad \Delta F''_2 = 0.279 ;$$
$$\Delta F''_1 = -1.202; \quad \Delta F''_2 = -0.622 . \tag{7.38}$$

In order to extract the anomalous scattering corrections for the considered reflection, the structure amplitude modulus should be measured with the accuracy better than $\delta \simeq 10\%$.

Let us now simulate the PXR characteristics for this reflection, considering the radiation from the electron beam with the energy $E = 30$ MeV and current $J = 1$ µA. The observation angles $\theta_{1,2}$, which allow us to measure the intensity of the radiation with two different frequencies, are calculated by means of formula (7.19):

$$\theta_{1,2} = \arcsin\left(\frac{gc^2}{2v\omega_{1,2}}\right); \qquad \theta_1 = 34.6°; \qquad \theta_2 = 24.6° . \tag{7.39}$$

These peaks are easily separated with necessary accuracy.

In some cases, it may be difficult to measure PXR intensities at two positions of the detector close to each other in the limits of the angular width of one reflection corresponding to the definite reciprocal lattice vector $\boldsymbol{g}$. In this situation, the measurements for the fixed frequency but for two different reflections corresponding to the reciprocal lattice vectors $\boldsymbol{g}$ and $(-\boldsymbol{g})$ can be performed. This approach was usually considered for ASM in the conventional diffraction experiments [38], and it is possible for the crystal with unit cell without the centre of symmetry ($\varphi_0 \neq 0; \pi$). In this case, a geometry of the PXR experiment should be inverted relatively to the electron velocity in comparison with the initial position and the amplitude $F(-\boldsymbol{g}, \omega_1)$ is calculated by means of the formula

$$F(-\boldsymbol{g},\omega_1) = |F_0(\boldsymbol{g})|(\cos\varphi_0 - \mathrm{i}\sin\varphi_0) + \Delta F'_{1,2} + \mathrm{i}\Delta F''_{1,2}\,. \tag{7.40}$$

Now we can use the values of the structure amplitudes (7.35–7.40) in formula (7.29) for calculation of the PXR intensities. In accordance with (7.37), for the reflection (113) the phase $\varphi_0$ is very close to zero because of the symmetry of the considered structure and, therefore, the use of the reflection $(\overline{113})$ is not effective. However, PXR intensities for two close frequencies (observation angles according to (7.39)) for the same reflection can be measured and the calculation results in the following values:

$$I_{\boldsymbol{g},\omega_1} = 1.28\times10^4\ [\mathrm{ph\ s^{-1}}];\qquad I_{\boldsymbol{g},\omega_2} = 1.08\times10^4\ [\mathrm{ph\ s^{-1}}]\,. \tag{7.41}$$

Here the value $L = L_{\mathrm{abs}} = 10\,\mu\mathrm{m}$ is used for the crystal thickness; $\theta_{\mathrm{D}} = 10^{-2}$ and the same values are used for the current and energy of the electron beam as in (7.38). The magnitude of the intensity is weakly sensitive to the accuracy of the definition of the experimental parameter $\theta_{\mathrm{D}}$, since it is included in the argument of the logarithmic function only. If the considered crystal is also characterized by some mosaicity parameter $\delta$, this parameter should be included in formula (7.29) along with the value $\theta_{\mathrm{D}}$ when calculating the PXR intensity [20]. Thus, the PXR intensities should be measured with the relative accuracy 15%. According to (7.34), such accuracy can be obtained for the observation time $t_{\mathrm{obs}} \simeq 20$ s if the photon intensity is given by (7.31).

For investigations of structures with the centre of symmetry it is important to distinguish whether the structure amplitude has the phase $\varphi_0 = 0$ or $\varphi_0 = \pi$. The considered approach could solve this problem as well. If the phase $\varphi_0$ in (7.38) is assumed to be equal to $\pi$ instead of zero, the correlation between the PXR intensities changes essentially:

$$I^{(-)}_{\boldsymbol{g},\omega_1} = 1.02\times10^4\ [\mathrm{ph\ s^{-1}}];\qquad I^{(-)}_{\boldsymbol{g},\omega_2} = 1.19\times10^4\ [\mathrm{ph\ s^{-1}}]\,. \tag{7.42}$$

In real experiment with the application of PXR for ASM, the main goal is to evaluate the phase $\varphi_0$, which means that the inverse problem should be solved. The modulus of the structure amplitudes has to be calculated on the basis of the intensity measurements for two different positions of the detector:

$$|F(\boldsymbol{g},\omega_{1,2})| = \sqrt{I_{\boldsymbol{g},\omega_{1,2}}A(E,\omega_{1,2})}\,, \tag{7.43}$$

with the function $A(E,\omega)$ defined by formula (7.31).

Then the unknown phase $\varphi_0$ can be calculated as the solution of the equations which follow from (7.38):

$$\begin{aligned}
\sin\varphi_0(\boldsymbol{g}) &= \frac{Q_1\Delta F'_2 - Q_2\Delta F'_1}{2|F_0(\boldsymbol{g})|P}\,;\\
\cos\varphi_0(\boldsymbol{g}) &= -\frac{Q_1\Delta F''_2 - Q_2\Delta F''_1}{2|F_0(\boldsymbol{g})|P}\,;\\
Q_{1,2} &= |F(\boldsymbol{g},\omega_{1,2})|^2 - |F_0(\boldsymbol{g})|^2 - (\Delta F'_1)^2 - (\Delta F''_1)^2\,;\\
P &= \Delta F''_1\Delta F'_2 - \Delta F''_2\Delta F'_1\,.
\end{aligned} \tag{7.44}$$

Both $\sin\varphi_0(\boldsymbol{g})$ and $\cos\varphi_0(\boldsymbol{g})$ should be calculated in order to find the sign of the phase uniquely [38]. The essential obstacle for practical realization of ASM on the basis of PXR is a radiation damage of the investigated crystal by an incident electron beam. If ASM is assumed to be used for studies of organic objects such as proteins, the probable radiation damage can be estimated in the following way. The expression [32] for energy loss by electrons of primary energy $E$ and current $J$ in unit time and within the layer of thickness $L$ is

$$\Delta E \approx J\frac{4\pi e^4 Z_{\mathrm{tot}} L}{\Omega m c^2} B(E)\,, \tag{7.45}$$

where $B(E)$ is a dimensionless function weakly dependent on $E$. The energy change $\Delta E$ causes the temperature increase $\Delta T$:

$$k_{\mathrm{B}}\frac{\zeta_{\mathrm{tot}} S L}{\Omega}\Delta T \approx \Delta E\,. \tag{7.46}$$

Here $k_{\mathrm{B}}$ is a Boltzmann constant, $k_{\mathrm{B}}\zeta_{\mathrm{tot}}$ is a specific heat of the crystallographic unit cell and $S$ is an electron beam cross-section. Using (7.45) and (7.46), a rough estimate for the crystal temperature increase in unit time can be derived:

$$\Delta T\ [\mathrm{K\ s^{-1}}] \approx \frac{Z_{\mathrm{eff}}}{\zeta_{\mathrm{eff}}}\frac{J\ [\mu\mathrm{A}]}{S\ [\mathrm{mm}^2]}\,, \tag{7.47}$$

where the averaged values $Z_{\mathrm{eff}}$ and $\zeta_{\mathrm{eff}}$ are normalized to a single atom of the crystallographic unit cell. Expression (7.47) can be used for prediction of radiation damages of the investigated samples. The practical values of this estimate have to be evaluated particularly for every possible ASM application.

In conclusion, the PXR in Mössbauer crystals (Sect. 2.4) can also be used in $\gamma$-resonant nuclear spectroscopy. Several applications of this set-up for correlation and temporal measurements have been discussed in [3,12].

## 7.3 Parametric Beam Instability and PXR Free-Electron Laser

From the point of view of quantum theory, the PXR is considered in the above section as a spontaneous radiation. In this section, we consider a possibility of observing induced PXR and developing an X-ray laser on this basis. Different mechanisms of the induced X-ray radiation are now actively studied, which may serve as a basis for construction of an X-ray laser, functioning over the 10–100 keV range. Various types of free-electron lasers, using the Compton scattering of a light wave by an electron beam [33], resonance transition radiation [36], channelling particle radiation [26,31], have been considered. The analysis of the obstacles revealed that large current densities $j > 10^{13}$ A cm$^{-2}$ of relativistic electrons are necessary to realize the induced radiation mode.

This fact puts a question mark on the possibility of the X-ray coherent generation in the mentioned range. The emission from charged particles in a crystal to large angles to the particle velocity has been shown [13–16] to provide a radical change in a gain of the radiation and reduces the required current density for the generation start to the values $j \sim 10^8$ A cm$^{-2}$, which make possible the development of PXR FEL. This possibility follows from the recently discovered law of parametric beam instability [13–16], which occurs when an electron (positron) beam passes through a crystal. This instability leads to a drastic increase of the coefficient of the induced radiation in the crystal and reduces the threshold current for the generation start. We consider here the classic theory of this phenomenon [9].

The classic approach to the electron movement assumes Maxwell's equations (2.1) for the description of the interaction between the charged particle and wave fields of radiation and the crystal. Keeping in mind the X-ray diffraction of emitted photons in the framework of two-wave approximation (A.1), these equations are

$$\begin{gathered}
\left(k^2 - \frac{\omega^2}{c^2}\epsilon_0\right) \boldsymbol{E}_{\boldsymbol{k},\omega} - \boldsymbol{k}(\boldsymbol{k}\boldsymbol{E}_{\boldsymbol{k},\omega}) - \frac{\omega^2}{c^2}\chi_{-g}\boldsymbol{E}_{\boldsymbol{g},\omega} = \frac{4\pi \mathrm{i}\omega}{c^2}\boldsymbol{j}(\boldsymbol{k},\omega) , \\
\left(k_g^2 - \frac{\omega^2}{c^2}\epsilon_0\right) \boldsymbol{E}_{\boldsymbol{g},\omega} - \boldsymbol{k}_g(\boldsymbol{k}_g\boldsymbol{E}_{\boldsymbol{g},\omega}) - \frac{\omega^2}{c^2}\chi_{g}\boldsymbol{E}_{\boldsymbol{k},\omega} = \frac{4\pi \mathrm{i}\omega}{c^2}\boldsymbol{j}(\boldsymbol{k}_g,\omega) , \\
\boldsymbol{k}(\epsilon_0\boldsymbol{E}_{\boldsymbol{k},\omega} + \chi_{-g}\boldsymbol{E}_{\boldsymbol{g},\omega}) = 4\pi \mathrm{i} n(\boldsymbol{k},\omega) , \\
\boldsymbol{k}_g(\epsilon_0\boldsymbol{E}_{\boldsymbol{g},\omega} + \chi_{g}\boldsymbol{E}_{\boldsymbol{k},\omega}) = 4\pi \mathrm{i} n(\boldsymbol{k}_g,\omega) , \\
\epsilon_0 = 1 + \chi_0, \;\; \omega n(\boldsymbol{k},\omega) + \boldsymbol{k}\boldsymbol{j}(\boldsymbol{k},\omega) = 0 . \qquad (7.48)
\end{gathered}$$

Here $\boldsymbol{E}_{\boldsymbol{k},\omega}$ is the amplitude of the direct wave and $\boldsymbol{E}_{\boldsymbol{g},\omega}$ is that of the diffracted wave with the vector $\boldsymbol{k}_g = \boldsymbol{k} + \boldsymbol{g}$, $n(\boldsymbol{k},\omega)$ and $\boldsymbol{j}(\boldsymbol{k},\omega)$ are the Fourier components of the charge density and current of the beam, respectively:

$$\begin{gathered}
n(\boldsymbol{k},\omega) = \int \mathrm{d}\boldsymbol{r}\mathrm{d}t \; \mathrm{e}^{\mathrm{i}(\omega t - \boldsymbol{k}\boldsymbol{r})} n(\boldsymbol{r},t) , \\
n(\boldsymbol{r},t) = e\sum_j \delta[\boldsymbol{r} - \boldsymbol{r}_j(t)], \qquad \boldsymbol{j}(\boldsymbol{r},t) = e\sum_j \boldsymbol{v}_j(t)\delta[\boldsymbol{r} - \boldsymbol{r}_j(t)] , \qquad (7.49)
\end{gathered}$$

for electrons on the trajectory $\boldsymbol{r}_j(t)$.

To analyse the influence of the diffraction on the instability, we separate the amplitudes of transverse $E_t^{(s)}$ (polarization $\boldsymbol{e}_{\mathrm{s}}$, $s = \sigma, \pi$) and longitudinal $E_l^{(s)}$ electromagnetic waves. The system of equations for these amplitudes is

$$\begin{gathered}
(k^2c^2 - \omega^2\epsilon_0)E_{\boldsymbol{k},t}^{(s)} - \omega^2\chi_{-g}C_{\mathrm{s}}E_{\boldsymbol{g},t}^{(s)} = 4\pi \mathrm{i}\omega j_t^{(s)}(\boldsymbol{k},\omega) , \\
(k_g^2c^2 - \omega^2\epsilon_0)E_{\boldsymbol{g},t}^{(s)} - \omega^2\chi_{g}C_{\mathrm{s}}E_{\boldsymbol{k},t}^{(s)} = 4\pi \mathrm{i}\omega j_t^{(s)}(\boldsymbol{k}_{\boldsymbol{g}},\omega) , \\
\omega E_l^{(s)}(\boldsymbol{k},\omega) = -4\pi \mathrm{i} j_l^{(s)}(\boldsymbol{k},\omega) . \qquad (7.50)
\end{gathered}$$

The longitudinal component also creates the wave with the charge density

$$4\pi n(\boldsymbol{k},\omega) = -kE_l^{(s)}(\boldsymbol{k},\omega) \; .$$

To obtain an explicit expression for the beam current, the coordinate $\boldsymbol{r}_j$ and velocity $\boldsymbol{v}_j(t)$ for the $j$th particle are represented as

$$\begin{aligned} \boldsymbol{v}_j(t) &= \boldsymbol{u} + \delta\boldsymbol{v}_{jt}(t) + \delta\boldsymbol{v}_{jl}(t) \; , \\ \boldsymbol{r}_j(t) &= \boldsymbol{r}_{j0} + \boldsymbol{u}t + \delta\boldsymbol{r}_j(t) + \delta\boldsymbol{v}_{jl}(t) \; , \end{aligned} \tag{7.51}$$

where $\boldsymbol{u}$ is the total beam velocity, $\delta\boldsymbol{v}_{jt}(t)$ and $\delta\boldsymbol{v}_{jl}(t)$ are perturbations of the velocity of the $j$th particle due to transverse and longitudinal waves, respectively, and $\delta\boldsymbol{r}_j(t)$ is the electron trajectory change due to interaction with the radiation wave field.

For the analysis of the beam instability, the system of equations has to be resolved within the linear approximation for amplitudes of the radiation field [18]. In this approximation, the Fourier components of transverse and longitudinal fields are

$$\begin{aligned} j_t^{(s)}(\boldsymbol{k},\omega) &= \mathrm{i}\frac{k(\boldsymbol{e}_{\mathrm{s}}\boldsymbol{u})}{4\pi}E_l^{(s)}(\boldsymbol{k},\omega) + e\sum_j \delta\boldsymbol{v}_{jt}(\omega - \boldsymbol{k}\boldsymbol{u})\mathrm{e}^{-\mathrm{i}\boldsymbol{k}\boldsymbol{r}_{j0}} \; , \\ j_l^{(s)}(\boldsymbol{k},\omega) &= \mathrm{i}\frac{(\boldsymbol{k}\boldsymbol{u})}{4\pi}E_l^{(s)}(\boldsymbol{k},\omega) + e\sum_j \delta\boldsymbol{v}_{jl}(\omega - \boldsymbol{k}\boldsymbol{u})\mathrm{e}^{-\mathrm{i}\boldsymbol{k}\boldsymbol{r}_{j0}} \; , \end{aligned} \tag{7.52}$$

where $\delta\boldsymbol{v}_{jt,l}(\omega)$ are obtained by the Fourier transformation of $\delta\boldsymbol{v}_{jt,l}(t)$. Maxwell's equations (7.48) have to be supplemented by equations of motion (2.1) for an electron in the electromagnetic field $E(\boldsymbol{r}_j(t),t)$:

$$\begin{aligned} \frac{\mathrm{d}\boldsymbol{v}_j(t)}{\mathrm{d}t} &= \frac{e}{m\gamma}E(\boldsymbol{r}_j(t),t) + \frac{e}{m\gamma c}(\boldsymbol{v}_j(t)\times H(\boldsymbol{r}_j(t),t)) \\ &\quad - \frac{e}{m\gamma c^2}v_j(t)v_j(t)E(\boldsymbol{r}_j(t),t) \; , \\ \frac{1}{c}\frac{\partial H(\boldsymbol{r}_j(t),t)}{\partial t} &= -\nabla\times E(\boldsymbol{r}_j(t),t) \; . \end{aligned} \tag{7.53}$$

Equations (7.53) are solved in linear approximation, and after Fourier transformation the longitudinal and transverse components of the velocity are ($\gamma^2 = 1 - u^2/c^2$)

$$\begin{aligned} -\mathrm{i}\omega\delta\boldsymbol{v}_{jt}^{(s)}(\omega) &= \frac{e}{m\gamma(2\pi)^3}\int \mathrm{d}\boldsymbol{k}'\mathrm{e}^{\mathrm{i}\boldsymbol{k}\boldsymbol{r}_{j0}} \\ &\times\left\{\left[1 - \frac{(\boldsymbol{e}_{\mathrm{s}}\boldsymbol{u})^2}{c^2}\right] - \frac{\boldsymbol{k}'\boldsymbol{u}}{\omega + \boldsymbol{k}'\boldsymbol{u}} - \frac{(\boldsymbol{e}_{\mathrm{s}}\boldsymbol{u})(\boldsymbol{k}\boldsymbol{u})}{kc^2}\right\}E_{\boldsymbol{k}',t}^{(s)}(\omega + \boldsymbol{k}'\boldsymbol{u}) \; , \\ -\mathrm{i}\omega\delta\boldsymbol{v}_{jl}^{(s)}(\omega) &= \frac{e}{m\gamma(2\pi)^3}\int \mathrm{d}\boldsymbol{k}'\mathrm{e}^{\mathrm{i}\boldsymbol{k}\boldsymbol{r}_{j0}} \\ &\times\left\{\left[1 - \frac{(\boldsymbol{k}\boldsymbol{u})^2}{k^2c^2}\right] + \left[\frac{k'(\boldsymbol{e}_{\mathrm{s}}\boldsymbol{u})}{\omega + \boldsymbol{k}'\boldsymbol{u}} - \frac{(\boldsymbol{k}'\boldsymbol{u})^2}{k'^2c^2}\right]\right\}E_{\boldsymbol{k}',t}^{(s)}(\omega + \boldsymbol{k}'\boldsymbol{u}) \; . \end{aligned} \tag{7.54}$$

For the primary homogeneous beam

$$\sum_j \mathrm{e}^{\mathrm{i}(\boldsymbol{k}-\boldsymbol{k}')\boldsymbol{r}_{j0}} = n_{\mathrm{B}}(2\pi)^3\delta(\boldsymbol{k}-\boldsymbol{k}'), \qquad n_{\mathrm{B}} = \frac{J}{ecS}\,, \tag{7.55}$$

where $n_{\mathrm{B}}$ is the electron density of the beam of current $J$ and cross-section $S$. Using (7.48)–(7.55), the following closed linear equation system for transverse and longitudinal wave fields is found for two-wave diffraction:

$$\begin{aligned}
\{k^2c^2 - \omega^2[\epsilon_0 + \chi_{\mathsf{tt}}^{(\mathsf{s})}(\boldsymbol{k})]\}E_{\boldsymbol{k},t}^{(s)} - \omega^2\chi_{-g}C_{\mathrm{s}}\boldsymbol{E}_{\boldsymbol{g},t}^{(s)} &= \omega^2\chi_{\mathsf{tl}}^{(\mathsf{s})}(\boldsymbol{k})E_{\boldsymbol{k},l}^{(s)}\,,\\
\{k_g^2c^2 - \omega^2[\epsilon_0 + \chi_{\mathsf{tt}}^{(\mathsf{s})}(\boldsymbol{k}_g)]\}E_{\boldsymbol{g},t}^{(s)} - \omega^2\chi_{g}C_{\mathrm{s}}\boldsymbol{E}_{\boldsymbol{g},t}^{(s)} &= \omega^2\chi_{\mathsf{tl}}^{(\mathsf{s})}(\boldsymbol{k}_g)E_{\boldsymbol{k}_g,l}^{(s)}\,,\\
\{(\omega - \boldsymbol{k}\boldsymbol{u})^2 - \omega^2\chi_{\mathsf{ll}}^{(\mathsf{s})}\}E_{\boldsymbol{k},l}^{(s)} &= \omega^2\chi_{\mathsf{lt}}^{(\mathsf{s})}(\boldsymbol{k})E_{\boldsymbol{k},l}^{(s)}\,.
\end{aligned} \tag{7.56}$$

$\chi_{\mathsf{tl}}^{(\mathsf{s})}$ is a tensor of the beam polarizability, which is physically analogous to the crystal polarizabilities:

$$\begin{aligned}
\chi_{\mathsf{tt}}^{(\mathsf{s})}(\boldsymbol{k}) &= -\frac{\omega_{\mathrm{B}}^2}{\gamma\omega^2}\left[1 - \frac{\omega(\boldsymbol{e}_{\mathrm{s}}\boldsymbol{u})^2}{c^2(\omega - \boldsymbol{k}\boldsymbol{u})}\right], \quad \chi_{\mathsf{tl}}^{(\mathsf{s})}(\boldsymbol{k}) = -\frac{\omega_{\mathrm{B}}^2}{\gamma\omega^3}k(\boldsymbol{e}_{\mathrm{s}}\boldsymbol{u})\,,\\
\chi_{\mathsf{ll}}^{(\mathsf{s})}(\boldsymbol{k}) &= \frac{\omega_{\mathrm{B}}^2}{\gamma\omega^2}\left[1 - \frac{(\boldsymbol{k}\boldsymbol{u})^2}{k^2c^2}\right], \quad \chi_{\mathsf{lt}}^{(\mathsf{s})}(\boldsymbol{k}) = \frac{\omega_{\mathrm{B}}^2}{\gamma\omega^2}(\boldsymbol{e}_{\mathrm{s}}\boldsymbol{u})\left[\frac{k}{\omega} - \frac{(\boldsymbol{k}\boldsymbol{u})}{kc^2}\right],\\
\omega_{\mathrm{B}}^2 &= \frac{4\pi e^2 n_{\mathrm{B}}}{m}\,.
\end{aligned} \tag{7.57}$$

The determinant of (7.57) delivers a dispersion equation for $k$ and $\omega$. For relativistic particles, the terms with $(\omega - \boldsymbol{k}\boldsymbol{u})$ in denominator are essential, whereas the terms with $(\omega - \boldsymbol{k}_g\boldsymbol{u})^{-1}$ are negligible. Thus, the dispersion equation is

$$\begin{aligned}
&\left\{\left[k^2c^2 - \omega^2\epsilon_0 + \frac{\omega_{\mathrm{B}}^2}{\gamma}\right]\left[k_g^2c^2 - \omega^2\epsilon_0 + \frac{\omega_{\mathrm{B}}^2}{\gamma}\right] - \omega^4C_{\mathrm{s}}^2\chi_g\chi_{-g}\right\} =\\
&-\frac{\omega_{\mathrm{B}}^2u^2}{\gamma(\omega - \boldsymbol{k}\boldsymbol{u})^2}\theta^2k\left(k - \frac{\omega u\cos\theta}{c^2}\right)\left[k_g^2c^2 - \omega^2\epsilon_0 + \frac{\omega_{\mathrm{B}}^2}{\gamma}\right] = \frac{A(k_z,\omega)}{(\omega - \boldsymbol{k}\boldsymbol{u})^2}\,.
\end{aligned} \tag{7.58}$$

Here $\theta \ll 1$ is an angle between $\boldsymbol{k}$ and the beam velocity $\boldsymbol{u}$; the latter corresponds to the axis $Z$. The most essential influence on the dispersion equation is exerted by variation of the current due to the calculation of the longitudinal component of the radiation field performed under the beam particles (the right-hand side of (7.58)). The solution of non-linear dispersion equation (7.58) is localized [18] in the intersection of dispersion equations, corresponding to the wave fields of radiation in the crystal and radiation of the spatial charged beam. The coordinates $k_{z0}$ and $\omega_0$ of this intersection are found from

$$\begin{aligned}
\{[k^2c^2 - \omega^2\epsilon_0][k_g^2c^2 - \omega^2\epsilon_0] - \omega^4C_{\mathrm{s}}^2\chi_g\chi_{-g}\} &\equiv D(k_z,\omega) = 0\,,\\
(\omega - \boldsymbol{k}\boldsymbol{u})^2 &= 0\,.
\end{aligned} \tag{7.59}$$

The frequency and wave vector of radiation have to satisfy the diffraction conditions at the time of PXR emission, and therefore

$$k_{z0} = k_{\mathrm{B}}(1+\delta), \quad \omega_0 = uk_{\mathrm{B}}(1+\delta), \quad |\delta| \ll 1\,, \tag{7.60}$$

with $k_{\mathrm{B}}$ determined from the Bragg condition for the excited reciprocal lattice vector:

$$k_{\mathrm{B}} = -\frac{2\boldsymbol{k}_{\perp}\boldsymbol{g}_{\perp} + g^2}{g \sin\theta_{\mathrm{B}}}, \quad \sin\theta_{\mathrm{B}} = \frac{(\boldsymbol{g}\boldsymbol{u})}{gu}\,, \tag{7.61}$$

and the deviation $\delta$ is

$$\delta = \frac{\chi_g \chi_{-g} - (\eta+\xi)^2}{2\nu(\eta+\xi)}\,,$$

$$\nu = -\frac{g\sin\theta_{\mathrm{B}}}{k_{\mathrm{B}}}, \qquad \eta = \frac{k_{\perp}^2}{k_{\mathrm{B}}^2} \approx \theta^2, \qquad \xi = \gamma^{-2} - \chi_0\,. \tag{7.62}$$

According to the general method for instability analysis [18, 33], the solution of the dispersion equation (7.58) may be found as an expansion series over the detunings $\omega'$ and $k'_z$ near the point (7.60):

$$k_z = k_{z0} + k'_z, \qquad \omega = \omega_0 + \omega', \qquad |k_{z0}| \gg |k'_z|, \qquad \omega_0 \gg |\omega'|\,;$$

$$\begin{aligned} D(k_z,\omega) \simeq\ & \omega' \frac{\partial D_0}{\partial \omega} + k'_z \frac{\partial D_0}{\partial k_z} + \frac{1}{2}\omega'^2 \frac{\partial^2 D_0}{\partial \omega^2} \\ & + \frac{1}{2}k_z'^2 \frac{\partial^2 D_0}{\partial k_z^2} + k_z \omega' \frac{\partial^2 D_0}{\partial k_z \partial \omega} + \ldots\,, \end{aligned} \tag{7.63}$$

where index 0 points to the coordinate $(k_{z0}, \omega_0)$.

For an arbitrary crystal orientation and radiation angle in (7.63), the first and second terms are enough to preserve an accuracy (such as for homogeneous medium [18]), and (7.58) is transformed to

$$(\omega' - uk'_z)^2(\omega' - v_g k'_z) = \frac{A(k_{z0},\omega_0)}{\partial D_0/\partial\omega}\,, \tag{7.64}$$

where the velocity of the spatial charge wave is

$$v_g = -\frac{\partial D_0/\partial k_z}{(\partial D_0)/\partial \omega} = \frac{c^2}{u\epsilon_0}\left[1 + \frac{\nu(\eta+\xi)^2}{\chi_g\chi_{-g} + (\eta+\xi)^2}\right]\,. \tag{7.65}$$

The most effective transmission of beam energy into radiation energy occurs under the synchronism condition [18] when the velocities of the wave and the beam are equal:

$$v_g = u\,. \tag{7.66}$$

As follows from (7.65), for X-rays this condition is fulfilled due solely to the crystal periodicity, i.e. at $\boldsymbol{g} \neq 0$, $\chi_g \neq 0$. From the instability theory [18, 33], the negative imaginary part $\mathrm{Im}(k_z)$, so-called increment of instability, is a quantitative criterion, which determines the exponential increase of primary perturbation. This parameter defines the amplification coefficient on the unit length, i.e. the gain

$$G\ [\mathrm{sm}^{-1}] = \mathrm{Im}(k_z)\ ,$$

which is the principal characteristics of FEL [33]. If (7.66) is satisfied, the solution of (7.64) results in the following $G$ for the parametric beam instability:

$$G = \frac{\sqrt{3}}{2}\left(\frac{Q\omega_0\chi_g\chi_{-g}}{u^2c^2k_{z0}[\chi_g\chi_{-g} + (1+\nu)(\eta+\xi)^2]}\right)^{1/3} ,$$
$$Q = -\frac{\omega_{\mathrm{B}}^2 u^2\theta^2 k_0(k_0 - \omega_0\cos\theta u/c^2)}{2\omega_0\gamma} . \tag{7.67}$$

Contrary to the Cherenkov instability in a homogeneous medium [1], the amplification coefficient in a 3D periodic medium depends essentially on the crystal material and orientation, which is a direct way to gain an interaction between the radiation and electron beam. The geometry of emission, when $\partial D_0/\partial k_z$ in (7.63) becomes zero, is of special interest [9, 14, 15]:

$$\theta = \left[\frac{|\chi_g|}{\sqrt{|\cos 2\theta_{\mathrm{B}}|}} - \frac{1}{\gamma^2} - |\chi_0|\right]^{1/2} ,$$
$$1 < 2\sin^2\theta_{\mathrm{B}} < 1 + \frac{\chi_g\chi_{-g}}{(\gamma^{-2}+|\chi_0|)^2} . \tag{7.68}$$

The diffracted wave in this case travels at an angle $\sim 90°$ to the beam velocity:

$$2\theta_{\mathrm{B}} = \frac{\pi}{2} + \psi\ ,$$
$$\theta\sin 2\theta_{\mathrm{B}} < \psi < \theta\sin 2\theta_{\mathrm{B}} + \frac{\chi_g\chi_{-g}}{(\gamma^{-2}+|\chi_0|)^2} . \tag{7.69}$$

In addition, to the longitudinal component, the transverse component of the diffracted wave becomes parallel to the electron velocity in this case and performs a supplementary work on beam modulation (bunching). Mathematically, this regime leads to the transformation of (7.64):

$$(\omega' - uk_z')^2(k_z'^2 - F\omega') = 2\frac{A(k_{z0},\omega_0)}{\partial^2 D_0/\partial k_z^2} ,$$
$$F = -2\frac{\partial D_0/\partial\omega}{\partial^2 D_0/\partial k_z^2} = -\frac{\epsilon_0 u^2|\chi_g| g\sin\theta_{\mathrm{B}}}{2c|\cos 2\theta_{\mathrm{B}}|^{3/2}} . \tag{7.70}$$

For the optimal case of small detuning $\omega' \to 0$, the gain of the parametric instability is

$$G=\left(\frac{Q\omega_0|\chi_g|}{|\cos 2\theta_{\mathrm{B}}|^{1/2}}\right)^{1/4} . \tag{7.71}$$

Thus, at certain choice of parameters of the 3D crystal, the dependence of the amplification coefficient on the particle density in the beam can qualitatively be changed, in comparison to the homogeneous medium: $n_{\mathrm{B}}^{1/3} \rightarrow n_{\mathrm{B}}^{1/4}$. Taking into consideration the inequality $\omega^3 Q \ll 1$, which is fulfilled for the current in real devices, the change in power degree influences significantly $G$. Later, Baryshevsky and Feranchuk [16] studied the influence of quantum effects on instability increment, and $N$-wave diffraction is shown to increase the increment due to the dependence

$$G \sim (n_{\mathrm{B}})^{1/(N+2)} .$$

The detailed theory and analysis of PXR FEL development on the basis of the parametric instability of charged particle beams passing through a crystal have been carried out in [7,8,10]. Under the conditions of multiwave diffraction, the generation threshold is proved to be reduced due to the phenomenon of the parametric beam instability. This fact makes it possible to observe an induced PXR radiation as well as an induced channelling radiation in a LiH crystal at the electron beam current density $j \sim 10^8$ cm$^{-2}$ and energy from tens to hundreds MeV. Moreover, this phenomenon is also observable [4] for the particles moving inside a slit made inside this crystal. This discovery extends the phenomenon of the parametric beam instability to the vacuum electronic devices, such as travelling wave tubes and backward wave oscillators, and to different types of spontaneous radiation. Based on these investigations, a new type of FEL has been proposed, named a volume free-electron laser (VFEL) [5]. The first lasing of VFEL in the millimetre wavelength range was observed in 2001 [6].

The advantage of FEL on the basis of undulator is the use of vacuum duct for an electron beam and radiation, which for the large undulator length $L_U$ results in a high integral amplification coefficient $K_U = G_U L_U$. For FEL, on the basis of PXR, the amplification length is limited by the absorption length $L_{\mathrm{abs}}$. However, the optimization of the parameters of the crystal or artificial periodic nanostructure [41,42,45] to increase the quantum output (Sect. 7.1) and the use of the diffraction geometry for anomalous transmission to increase the $L_{\mathrm{abs}}$ (Sect. 3.4) both allow us to increase the integral amplification coefficient for PXR, too.

## References

1. A.I. Akhiezer, Ya.B. Fainberg: Dokl. Akad. Nauk USSR **69**, 555 (1949)
2. M.Yu. Andreyashkin, V.V. Kaplin, S.R. Uglov, V.N. Zabaev, M. Piestrup: Appl. Phys. Lett. **72**, 1385 (1998)

3. V.G. Baryshevsky: *Channelling, Radiation and Reactions at Crystals under High Energy* (Belarussian State University, Minsk 1982)
4. V.G. Baryshevsky: Dokl. Akad. Sci. USSR **299**, 19 (1988)
5. V.G. Baryshevsky: Nucl. Instrum. Methods A **445**, 281 (2000)
6. V.G. Baryshevsky, K.G. Batrakov, A.A. Gurinovich, I.I. Ilienko, A.S. Lobko, V.I. Moroz, P.F. Sofronov, V.I. Stolyarsky: Nucl. Instrum. Methods A **483**, 21 (2002)
7. V.G. Baryshevsky, K.G. Batrakov, I.Ya. Dubovskaya: J. Phys. D: Appl. Phys. **24**, 1250 (1991)
8. V.G. Baryshevsky, K.G. Batrakov, I.Ya. Dubovskaya: Nucl. Instrum. Methods A **358**, 93 (1995); Nucl. Instrum. Methods A **375**, 292 (1996)
9. V.G. Baryshevsky, I.Ya. Dubovskaya, I.D. Feranchuk: Izv. Belarussian Akad. Nauk, Ser. Fiz-Mat. Nauk **1**, 92 (1988)
10. V.G. Baryshevsky, I.Ya. Dubovskaya, A.V. Zege: Phy. Lett. A **149**, 30 (1990); Nucl. Instrum. Methods B **51**, 368 (1990)
11. V.G. Baryshevsky, I.D. Feranchuk: Izv. Belarusian Acad. Sci. (Ser. Fiz.-Mat. Nauk) **N 2**, 117 (1975)
12. V.G. Baryshevsky, I.D. Feranchuk: Phys. Lett. A **76**, 452 (1980)
13. V.G. Baryshevsky, I.D. Feranchuk: Dokl. Belarussian Akad. Nauk **27**, 995 (1983)
14. V.G. Baryshevsky, I.D. Feranchuk: Dokl. Belarussian Akad. Nauk **28**, 336 (1984)
15. V.G. Baryshevsky, I.D. Feranchuk: Phys. Lett. **A 102**, 141 (1984)
16. V.G. Baryshevsky, I.D. Feranchuk: Izv. Belarussian Akad. Nauk, Ser. Fiz-Mat. Nauk **2**, 79 (1985)
17. Y. Danon, B.A. Sones, R.C. Block: Novel X-ray source at the RPI LINAC *ANS Transactions, Summer Meeting,* Hollywood, FL, 2002, vol **86**; B. Sones, Y. Danon, R. Block: Advances in parametric X-ray production at the RPI linear accelerator, *ANS Annual Meeting, ANS Transactions*, San Diego, CA vol **88** (2003) p. 352
18. N.S. Erokhin, M.V. Kuzelev, S.S. Moiseev, A.A. Rukhadze, A.B. Swartzburg: *Non-Equilibrium and Resonant Processes in Plasma Radiophysics* (Nauka, Moscow 1982)
19. I.D. Feranchuk, L.I. Gurskii, L.I. Komarov, O.M. Lugovskaya, F. Burgäzy, A.P. Ulyanenkov: Acta Crystallogr. A **58**, 370 (2002)
20. I.D. Feranchuk, A.V. Ivashin: J. Phys. (Paris) **46**, 1981 (1985)
21. I.D. Feranchuk, A.V. Ivashin: Krystallographiya **34**, 39 (1989)
22. I.D. Feranchuk, A.P. Ulyanenkov: Phys. Rev. B **63**, 155318 (2001)
23. I.D. Feranchuk, A.P. Ulyanenkov: Phys. Rev. B **68**, 235307 (2003)
24. I.D. Feranchuk, A.P. Ulyanenkov: Acta Cryst. A **61**, 125 (2005)
25. I.D. Feranchuk, A.P. Ulyanenkov, J. Harada, J.C.H. Spence, Phys. Rev. E **62**, 4225 (2000)
26. A. Friedman, A. Gover, G. Kurizki, S. Puschin, A. Yariv: Rev. Mod. Phys. **60**, 471 (1988)
27. G.V. Gurskaya, V.E. Zavodnik, A.D. Shutalev: Krystallographiya **48**, 461 (2003)
28. J. Harada, S. Hayashi, K. Omote, V.G. Baryshevsky, I.D. Feranchuk, A.P. Ulyanenkov: *X-ray Generating Method and Device, Patent HO6G 2/00, HO1J 35/08* (Japanese Patent Office, 2000)
29. M. Hart: Nucl. Instrum. Methods **A 297**, 306 (1990)
30. V.V. Kaplin, S.R. Uglov, V.N. Zabaev, M.A. Piestrup, C.K. Gary, N.N. Nasonov, M.K. Fuller: Appl. Phys. Lett. **76**, 3647 (2000)

31. G. Kurizki, M. Strauss, I. Oreg, N. Rostoker: Phys. Rev. A **35**, 3427 (1987)
32. L.D. Landau, E.M. Lifshitz: *Electrodynamics of Continuous Media* (Nauka, Moscow 1982)
33. T.C. Marshall: *Free-Electron Lasers* (Macmillan, New York, London 1985)
34. V.E. Pafomov: Zh. Tekhn. Fiz **33**, 557 (1963) (Sov. Phys. Tech. Phys. **8**, 412 (1963)); Proc. Lebedev Inst. Phys. **44**, 25 (1971)
35. M.A. Piestrup, L.W. Lombardo, J.T. Cremer, G.A. Retzlaff, L.M. Silzer, D.M. Skopik, V.V. Kaplin: unpublished
36. M.A. Piestrup, R.L. Finman: IEEE J. Quantum Electron. **19**, 357 (1988)
37. M.A. Piestrup, X. Wu, V.V. Kaplin, S.R. Uglov, J.T. Cremer, D.W. Rule, R.B. Fiorito: Rev. Sci. Instrum. **72**, 2159 (2001)
38. M.A. Porai-Koshitz: *Practical Course for X-ray Structure Analysis, vol 2* (Fizmatgiz, Moscow 1968)
39. B. Sones, Y. Danon, R.C. Block: Nucl. Instrum. Methods B **227**, 22 (2005)
40. B. Sones, Y. Danon, R. Block: Novel X-ray imaging opportunities for the RPI linear accelerator's tunable, quasi-monochromatic X-ray source *ANS Annual Meeting* (Pittsburgh, PA, 2004)
41. A. Ulyanenkov, T. Baumbach, N. Darowski, U. Pietsch, K.H. Wang, A. Forchel, T. Wiebach: J. Appl. Phys. **85** 1524 (1999)
42. A. Ulyanenkov, N. Darowski, J. Grenzer, U. Pietsch, K.H. Wang, A. Forchel: Phys. Rev. B **60** 16701 (1999)
43. A. Ulyanenkov, I.D. Feranchuk, L.I. Komarov: 50th DXC Proceedings, Steamboat Springs, p 55 (2001)
44. A. Ulyanenkov, I.D. Feranchuk, J. Harada: XTOP Proceedings, Poland, p 103 (2000)
45. A. Ulyanenkov, K. Inaba, P. Mikulik, N. Darowski, K. Omote, U. Pietsch, J. Grenzer, A. Forchel: J. Phys. D: Appl. Phys. **3**4 A179 (2001)
46. B.K. Vainshtein: *Modern Crystallography*, vol 1 (Nauka, Moscow 1978)
47. M.M. Woolfson and H.F. Fan: *Physical and Non-Physical Methods of Solving Crystal Structures* (Cambridge University Press, England 1995)
48. *X-ray Data Booklet* (Lowrence Berkeley National Laboratory, Berkeley 2001)

# A

# Appendix

## A.1 X-ray Polarizability and Eigenwaves for the Electromagnetic Field in a Crystal

The general expression for the tensor of the dielectric permittivity $\epsilon_{\mathrm{ij}}(\boldsymbol{k}, \boldsymbol{k}_g, \omega)$ in the constitutive equation (2.4) contains the tensor of the X-ray polarizability $\chi_{\mathrm{ij}}$, which describes the interaction of X-ray radiation with the crystal:

$$\epsilon_{\mathrm{ij}}(\boldsymbol{k}, \boldsymbol{k}_g, \omega) = \delta_{ij}\delta_{\boldsymbol{k},\boldsymbol{k}_g} + \chi_{\mathrm{ij}}(\boldsymbol{k}, \boldsymbol{k}_g, \omega), \qquad i, j = 1, 2, 3 \; . \tag{A.1}$$

The components of $\chi_{\mathrm{ij}}$ are not phenomenological parameters but microscopic characteristics of the crystal, which are expressed through the amplitudes of the scattering of X-ray photons on periodically arranged atoms and nuclei (see, for example, [6]):

$$\chi_{\mathrm{ij}}(\boldsymbol{k}, \boldsymbol{k}_g, \omega) = \frac{4\pi c^2}{\omega^2 \Omega} \sum_a \left[ f_{ij,a}^{(e)}(\boldsymbol{g}, \omega) + f_{ij,a}^{(n)}(\boldsymbol{k}, \boldsymbol{k}_g, \omega) \right] \mathrm{e}^{\mathrm{i}\boldsymbol{g}\boldsymbol{R}_a} \; . \tag{A.2}$$

Here $\Omega$ is the volume of a crystallographic unit cell; $R_a$ is the coordinate of the $a$th atom in the cell; $f_{ij,a}^{(e)}$ is the amplitude of the elastic coherent scattering of photons on atom's electrons [8]:

$$f_{ij,a}^{(e)}(\boldsymbol{g}, \omega) = -\delta_{ij} r_0 [F_a(\boldsymbol{g}) + \Delta f'(\omega) + \mathrm{i}\Delta f''(\omega)] \mathrm{e}^{-W_a(\boldsymbol{g})} \; , \tag{A.3}$$

where $r_0 = e^2/mc^2$ is the electromagnetic radius of an electron; $F_a(\boldsymbol{g})$ is an atomic scattering factor; $\Delta f_a'(\omega)$, $\Delta f_a''(\omega)$ are the real and imaginary parts of anomalous dispersion corrections, respectively, which take into account the absorption and resonant scattering of photons; $\mathrm{e}^{-W_a(\boldsymbol{g})}$ is the Debye–Waller factor, which quantifies the reduction of the elastic amplitude due to inelastic scattering on the crystal phonons. The method for calculation of the X-ray polarizability for various crystals is presented in [7].

The contribution to polarizability by the scattering of photons on resonant nuclear transitions is essential for Mössbauer crystals [1]:

$$f_{ij,a}^{(n)}(\boldsymbol{k},\boldsymbol{k}_g,\omega) = -\frac{\tilde{n}}{4\omega_\mathrm{r}}\frac{2J+1}{2J_0+1}\frac{\Gamma_1}{\hbar(\omega-\omega_\mathrm{r})+\Gamma}P_{ij}(\boldsymbol{k},\boldsymbol{k}_g)\eta_a \mathrm{e}^{-W_a(\boldsymbol{k},\boldsymbol{k}_g)}. \quad \text{(A.4)}$$

where $\omega_\mathrm{r}$ is the frequency of resonant transition of nuclei of the cell with weight $\eta_a$; $\Gamma_1$ and $\Gamma$ are the elastic and total widths of the excited level, respectively; $J_1$ and $J$ are the angular moments of the excited and ground states, respectively; the polarization factor $P_{ij}(\boldsymbol{k},\boldsymbol{k}_g)$ is determined by the multiplicity of transition:

$$E1 \to P_{ij} = \delta_{ij}, \quad M1 \to P_{ij} = \frac{(\boldsymbol{k}\boldsymbol{k}_g)\delta_{ij} - k_i k_{gj}}{k^2},$$

$$E2 \to P_{ij} = \frac{1}{k^2}[(\boldsymbol{k}\boldsymbol{k}_g)\delta_{ij} + k_j k_{gi} - 2k_i k_{gj}].$$

The Debye–Waller factor depends on the ratio of the width $\Gamma$ to the phonon energy of a crystal $\hbar\omega_\mathrm{phon}$:

$$W_a(\boldsymbol{k},\boldsymbol{k}_g) = \frac{1}{2}\overline{u_a^2}(k^2+k_g^2), \qquad \Gamma \ll \hbar\omega_\mathrm{phon},$$

$$W_a(\boldsymbol{k},\boldsymbol{k}_g) = \frac{1}{2}\overline{u_a^2}g^2, \qquad \Gamma \gg \hbar\omega_\mathrm{phon},$$

where $\overline{u_a^2}$ is the mean square amplitude of nuclear oscillations near an equilibrium position.

In the X-ray domain, the X-ray polarizability is typically $|\chi_\mathrm{i,j}| \sim 10^{-4}$–$10^{-6}$. For the solution of Maxwell's equations (2.1) with the accuracy $O(|\chi_\mathrm{i,j}|^2)$, the electromagnetic field in a medium remains transverse, and the interaction between the field and the crystal is essential at the wave vectors $\boldsymbol{k}$, satisfying the Bragg condition [9]:

$$\alpha_\mathrm{B} = \frac{2\boldsymbol{k}\boldsymbol{g}+g^2}{k^2} \leq \chi_0\,. \qquad \text{(A.5)}$$

In most cases, condition (A.5) is fulfilled for only one reciprocal lattice vector $\boldsymbol{g}$ for fixed $\boldsymbol{k}$, and the two-wave approximation of the dynamical diffraction theory is valid [3]. Then, the eigenwaves of the electromagnetic field, required for description of processes in the crystal (Sect. 3.2), are composed of the linear combination of plane waves:

$$\boldsymbol{A}_{\boldsymbol{k}\omega}^{(s)}(\boldsymbol{r}) = \boldsymbol{e}_s A_{\boldsymbol{k}s}\mathrm{e}^{\mathrm{i}\boldsymbol{k}\boldsymbol{r}} + \boldsymbol{e}_{gs}A_{gs}\mathrm{e}^{\mathrm{i}\boldsymbol{k}_g\boldsymbol{r}}, \qquad s=\sigma,\pi,$$

$$\boldsymbol{E}_{\boldsymbol{k}\omega}^{(s)}(\boldsymbol{r}) = \mathrm{i}\frac{\omega}{c}\boldsymbol{A}_{\boldsymbol{k}\omega}^{(s)}(\boldsymbol{r}), \qquad \text{(A.6)}$$

where the unit polarization vectors are

$$\boldsymbol{e}_\sigma \parallel \boldsymbol{e}_{g\sigma} \parallel [k\times g], \qquad \boldsymbol{e}_\pi \parallel [k\times[k\times g]], \qquad \boldsymbol{e}_{g\pi} \parallel [k_g\times[k\times g]],$$

and the amplitudes of wave (A.6) have to satisfy the algebraic equations:

$$\{k^2 - k_0^2(1+\chi_{00}^{(s)})\}A_{ks} - k_0^2\chi_{01}^{(s)}A_{gs} = 0\,,$$
$$\{k_g^2 - k_0^2(1+\chi_{11}^{(s)})\}A_{gs} - k_0^2\chi_{10}^{(s)}A_{ks} = 0\,,$$
$$k_0 = \frac{\omega}{c}, \qquad \chi_{00}^{(s)} = e_{si}e_{sj}\chi_{\rm ij}(\boldsymbol{k},\boldsymbol{k}), \quad \chi_{11}^{(s)} = e_{gsi}e_{gsj}\chi_{\rm ij}(\boldsymbol{k}_g,\boldsymbol{k}_g)\,,$$
$$\chi_{01}^{(s)} = e_{si}e_{gsj}\chi_{\rm ij}(\boldsymbol{k},\boldsymbol{k}_g), \qquad \chi_{10}^{(s)} = e_{gsi}e_{sj}\chi_{\rm ij}(\boldsymbol{k}_g,\boldsymbol{k})\,. \tag{A.7}$$

If the resonant scattering of X-rays on atoms and nuclei is negligible, then

$$\chi_{00}^{(s)}) = \chi_{11}^{(s)}) = \chi_0, \qquad \chi_{01}^{(s)} = \chi_{-g}C_{\rm s}, \quad \chi_{10}^{(s)} = \chi_g C_{\rm s},$$
$$C_\sigma = 1, \qquad C_\pi = \cos 2\theta_{\rm B}\,.$$

The condition of zero determinant for system (A.7) delivers the effective refraction indices for eigenwaves inside a crystal:

$$\boldsymbol{k}_{\mu s} = k_0\frac{\boldsymbol{k}}{k}n_{\mu s}, \; n_{\mu s}(1+\epsilon_{\mu 1}), \qquad \mu = 1,2\,,$$
$$\epsilon_{\mu s} = \frac{1}{4}\left[q \pm \sqrt{q^2 + 4\beta\chi_{00}\alpha_{\rm B} - \chi_{00}\chi_{11} + \chi_{01}^{(s)}\chi_{10}^{(s)}}\right]\,,$$
$$q = \chi_{00} + \beta\chi_{11} - \beta\alpha_{\rm B}, \; \beta = \frac{\gamma_0}{\gamma_g}, \; \gamma_0 = \cos(\boldsymbol{k},\boldsymbol{N}), \; \gamma_g = \cos(\boldsymbol{k}_g,\boldsymbol{N}), \tag{A.8}$$

where $\boldsymbol{N}$ is a normal to the crystal surface.

## A.2 Asymptotic for the Green Function and Boundary Conditions for the Electromagnetic Field

The asymptotic of the Green function (Sect. 2.1) for Maxwell's equations in the medium with an arbitrary dielectric permittivity is derived here in the limit $r \gg r'$. This function is the solution of the following equation ($k_0 = \omega/c, \; \alpha,\beta = 1,2,3$):

$$\varepsilon_{\alpha\beta\gamma}\varepsilon_{\gamma\mu\nu}\frac{\partial^2}{\partial x_\beta \partial x_\mu}G_{\nu\lambda}(\boldsymbol{r},\boldsymbol{r}',\omega) - \; k_0^2\int \mathrm{d}\boldsymbol{r}_1\epsilon_{\alpha\beta}(\boldsymbol{r},\boldsymbol{r}_1,\omega)G_{\beta\lambda}(\boldsymbol{r}_1,\boldsymbol{r}',\omega)$$
$$= \delta_{\alpha\lambda}\delta(\boldsymbol{r}-\boldsymbol{r}')\,. \tag{A.9}$$

Expressing the dielectric permittivity through the X-ray polarizability

$$\epsilon_{\alpha\beta}(\boldsymbol{r},\boldsymbol{r}_1,\omega) = \delta_{\alpha\beta}\delta(\boldsymbol{r}-\boldsymbol{r}_1) + \chi_{\alpha\beta}(\boldsymbol{r},\boldsymbol{r}_1,\omega)\,,$$

Equation (A.9) is reformulated in the integral form:

$$G_{\alpha\beta}(\boldsymbol{r},\boldsymbol{r}',\omega) = G_{\alpha\beta}^{(0)}(\boldsymbol{r},\boldsymbol{r}',\omega)$$
$$+ \; k_0^2\int \mathrm{d}\boldsymbol{r}_1\mathrm{d}\boldsymbol{r}_2 G_{\alpha\mu}^{(0)}(\boldsymbol{r},\boldsymbol{r}_1,\omega)\chi_{\mu\nu}(\boldsymbol{r}_1,\boldsymbol{r}_2,\omega)G_{\nu\beta}(\boldsymbol{r}_2,\boldsymbol{r}'\omega)\,. \tag{A.10}$$

The Green function $G^{(0)}_{\alpha\beta}(\boldsymbol{r},\boldsymbol{r}',\omega)$ for equation (A.9) in vacuum is represented [12] as an expansion of eigenstates of the free electromagnetic field:

$$G^{(0)}_{\alpha\beta}(\boldsymbol{r},\boldsymbol{r}',\omega) = \frac{1}{(2\pi)^3}\int \mathrm{d}\boldsymbol{q} \sum_{s=1,2} \frac{e^{(s)}_{\alpha}(\boldsymbol{q})e^{(s)*}_{\beta}(\boldsymbol{q})\mathrm{e}^{\mathrm{i}\boldsymbol{q}(\boldsymbol{r}-\boldsymbol{r}')}}{q^2 - k_0^2\,(1+iO)}\,, \qquad \text{(A.11)}$$

where $\boldsymbol{e}^{(s)}(\boldsymbol{q}),\ s=1,2;\ (\boldsymbol{q},\boldsymbol{e}^{(s)})=0$ are two mutually orthogonal unit vectors of polarization of the plane electromagnetic wave with the wave vector $\boldsymbol{q}$. In the considered case here, $r \gg r'$, the following asymptotic for the function $G^{(0)}_{\alpha\beta}$ is valid:

$$G^{(0)}_{\alpha\beta}(\boldsymbol{r},\boldsymbol{r}',\omega) \approx \frac{1}{4\pi}\sum_{s=1,2} e^{(s)}_{\alpha}(\boldsymbol{k})e^{(s)*}_{\beta}(\boldsymbol{k})\frac{\mathrm{e}^{\mathrm{i}k_0 r}}{r}\mathrm{e}^{-\mathrm{i}\boldsymbol{k}\boldsymbol{r}'}\,, \qquad \boldsymbol{k} = k_0\frac{\boldsymbol{r}}{r}\,. \qquad \text{(A.12)}$$

Using (A.12), the asymptotic (A.10) can be written as

$$G_{\alpha\beta}(\boldsymbol{r},\boldsymbol{r}',\omega) \approx \frac{1}{4\pi}\frac{\mathrm{e}^{\mathrm{i}k_0 r}}{r}\sum_{s=1,2} e^{(s)}_{\alpha}(\boldsymbol{k})$$
$$\times\left[e^{(s)}_{\beta}(\boldsymbol{k})\mathrm{e}^{\mathrm{i}\boldsymbol{k}\boldsymbol{r}'} + k_0^2\int \mathrm{d}\boldsymbol{r}_1\mathrm{d}\boldsymbol{r}_2\mathrm{e}^{\mathrm{i}\boldsymbol{k}\boldsymbol{r}_1}\chi^*_{\mu\nu}(\boldsymbol{r}_1,\boldsymbol{r}_2,\omega)G^*_{\nu\beta}(\boldsymbol{r}_2,\boldsymbol{r}'\omega)\right]^*. \qquad \text{(A.13)}$$

If the iterative solution of (A.10) for the exact Green function is used, the expression in the square brackets in (A.13) is represented as the series

$$E^{(s,-)}_{\boldsymbol{k},\beta}(\boldsymbol{r}) = e^{(s)}_{\beta}(\boldsymbol{k})\mathrm{e}^{\mathrm{i}\boldsymbol{k}\boldsymbol{r}}$$
$$+\,k_0^2\int \mathrm{d}\boldsymbol{r}_1\mathrm{d}\boldsymbol{r}_2 G^{(0,*)}_{\nu\mu}(\boldsymbol{r},\boldsymbol{r}_1\omega)\chi^*_{\beta\nu}(\boldsymbol{r}_1,\boldsymbol{r}_2,\omega)e^{(s)}_{\mu}\mathrm{e}^{\mathrm{i}\boldsymbol{k}\boldsymbol{r}_2} + \ldots\,, \qquad \text{(A.14)}$$

where the asymptotic behaviour of the function

$$G^{(0,*)}_{\alpha\beta}(\boldsymbol{r},\boldsymbol{r}',\omega) \approx \frac{1}{4\pi}\sum_{s=1,2} e^{(s)*}_{\alpha}(\boldsymbol{k})e^{(s)}_{\beta}(\boldsymbol{k})\frac{\mathrm{e}^{-\mathrm{i}k_0 r}}{r}\mathrm{e}^{\mathrm{i}\boldsymbol{k}\boldsymbol{r}'}$$

corresponds to a convergent spherical wave.

We now consider the eigenstates of the electromagnetic field in the medium with the dielectric permittivity $\epsilon^*_{\alpha\beta}(\boldsymbol{r},\boldsymbol{r}_1,\omega)$, which are the solutions of the homogeneous equation analogous to (A.9):

$$\varepsilon_{\alpha\beta\nu}\varepsilon_{\nu\mu\gamma}\frac{\partial^2}{\partial x_\beta\partial x_\mu}E^{(s,-)}_{\boldsymbol{k},\gamma} - k_0^2\int \mathrm{d}\boldsymbol{r}_1\epsilon^*_{\alpha\beta}(\boldsymbol{r},\boldsymbol{r}_1,\omega)E^{(s,-)}_{\boldsymbol{k},\beta} = 0\,. \qquad \text{(A.15)}$$

The integral form of this equation, after using the Green function $G^{(0,*)}_{\alpha\beta}(\boldsymbol{r},\boldsymbol{r}',\omega)$ and normalizing to the unit amplitude of the incident wave, is

$$E_{\boldsymbol{k},\alpha}^{(s,-)}(\boldsymbol{r}) = e_\alpha^{(s)}(\boldsymbol{k})\mathrm{e}^{\mathrm{i}\boldsymbol{kr}} + k_0^2 \int \mathrm{d}\boldsymbol{r}_1 \mathrm{d}\boldsymbol{r}_2 G_{\alpha\beta}^{(0,*)}(\boldsymbol{r},\boldsymbol{r}_1,\omega)\chi_{\beta\gamma}^*(\boldsymbol{r}_1,\boldsymbol{r}_2,\omega)E_{\boldsymbol{k},\beta}^{(s,-)}(\boldsymbol{r}_2) = 0\,, \quad \text{(A.16)}$$

and the iterative solution of this equation is delivered by series (A.15), which confirms (2.11) in Sect. 2.1.

The matrix elements for amplitudes of the electromagnetic field in the crystal are calculated (see Sect. 3.1) on the basis of the vector potential $\boldsymbol{A}_{\boldsymbol{k}}^{(s,-)}(\boldsymbol{r})$ and boundary conditions at the entrance and exit surfaces of the sample. The conventional boundary conditions of electrodynamics (the continuity of the tangential component of the field strength vectors and the normal components of the induction vectors [10]) in the X-ray region are reduced [9] with the accuracy $O(|\chi_0|^2)$ to the continuity of all field components and their derivatives. Within the framework of the two-wave approximation for the dynamical diffraction theory, (A.6)–(A.8) have to be taken into account. Then, the normalized vector potential, which is continuous at the sample surfaces $z = 0$ and $z = L$, is (see 3.14)

$$\boldsymbol{A}_{\boldsymbol{k}}^{(s,-)}(\boldsymbol{r}) = \sqrt{4\pi}\mathrm{e}^{\mathrm{i}\boldsymbol{kr}}\{\boldsymbol{e}_\mathrm{s}\Phi^{(s)}(z) + \boldsymbol{e}_{gs}\Phi_g^{(s)}(z)\mathrm{e}^{\mathrm{i}\boldsymbol{gr}}\}\,. \quad \text{(A.17)}$$

The explicit expressions for $\Phi^{(s)}(z)$ and $\Phi_g^{(s)}(z)$ in the case of different diffraction geometries and photon observation angles are as follows:

(1a) The Bragg case ($\beta < 0,\ \gamma_0 > 0,\ \gamma_g < 0$):

$$\begin{aligned}
\Phi^{(s)}(z) &= \{D_{s0}^*(0)H(-z) + D_{s0}^*(z)H(z)H(L-z) + H(z-L)\}\,\mathrm{e}^{-\mathrm{i}k_zL}\,,\\
\Phi_g^{(s)}(z) &= \beta\xi_\mathrm{s}^{g*}\left\{D_{sg}^*(z)H(z)H(L-z) + D_{sg}^*(L)H(z-L)\right\}\mathrm{e}^{-\mathrm{i}k_zL}\,,\\
D_{s0}(z) &= \xi_{1s}^0\mathrm{e}^{-\mathrm{i}k_0\epsilon_{1s}z/\gamma_0} + \xi_{2s}^0\mathrm{e}^{-\mathrm{i}k_0\epsilon_{2s}z/\gamma_0}\,,\\
D_{sg}(z) &= \mathrm{e}^{-\mathrm{i}k_0\epsilon_{1s}z/\gamma_0} - \mathrm{e}^{-\mathrm{i}k_0\epsilon_{2s}z/\gamma_0}\,,\\
\xi_{1,2s}^0 &= \pm\frac{2\epsilon_{2,1s} - \chi_{00}}{\Delta_\mathrm{s}}\,, \qquad \xi_\mathrm{s}^g = \frac{\chi_{10}^s}{\Delta_\mathrm{s}}\,,\\
\Delta_\mathrm{s} &= (2\epsilon_{2s} - \chi_{00})\mathrm{e}^{-\mathrm{i}k_0\epsilon_{1s}z/\gamma_0} - (2\epsilon_{1s} - \chi_{00})\mathrm{e}^{-\mathrm{i}k_0\epsilon_{2s}z/\gamma_0}\,.
\end{aligned} \quad \text{(A.18)}$$

(1b) The Bragg case ($\beta < 0,\ \gamma_0 < 0,\ \gamma_g > 0$):

$$\begin{aligned}
\Phi^{(s)}(z) &= H(-z) + D_{s0}^{(1*)}(z)H(z)H(L-z) + D_{s1}^{(1*)}(L)H(z-L)\,,\\
\Phi_g^{(s)}(z) &= -\beta\xi_\mathrm{s}^{g*}\left\{D_{sg}^{(1*)}(0)H(-z) + D_{sg}^{(1*)}(z)H(z)H(L-z)\right\}\,,\\
D_{s0}^{(1)}(z) &= \xi_{1s}^0\mathrm{e}^{\mathrm{i}k_0(\epsilon_{1s}z+\epsilon_{2s}L)/|\gamma_0|} + \xi_{2s}^0\mathrm{e}^{\mathrm{i}k_0(\epsilon_{2s}z+\epsilon_{1s}L)/|\gamma_0|}\,,\\
D_{sg}^{(1)}(z) &= \mathrm{e}^{\mathrm{i}k_0(\epsilon_{1s}z+\epsilon_{2s}L)/|\gamma_0|} - \mathrm{e}^{\mathrm{i}k_0(\epsilon_{2s}z+\epsilon_{1s}L)/|\gamma_0|}\,.
\end{aligned} \quad \text{(A.19)}$$

(2a) The Laue case ($\beta > 0,\ \gamma_0 > 0,\ \gamma_g > 0$):

$$\begin{aligned}\Phi^{(s)}(z) &= D_{s0}^{(2*)}(L)H(-z) + D_{s0}^{(2*)}(L-z)H(z)H(L-z) \\ &\quad + \mathrm{e}^{-\mathrm{i}k_z L}H(z-L)\,, \\ \Phi_g^{(s)}(z) &= \beta\left\{D_{sg}^{(2*)}(L)H(-z) + D_{sg}^{(2*)}(L-z)H(z)H(L-z)\right\}\,, \\ D_{s0}^{(2)}(z) &= -\zeta_{1s}^{0}\mathrm{e}^{\mathrm{i}k_0\epsilon_{1s}z/\gamma_0} - \zeta_{2s}^{0}\mathrm{e}^{\mathrm{i}k_0\epsilon_{2s}z/\gamma_0}]\,, \\ D_{sg}^{(2)}(z) &= \zeta_{1s}^{g}\mathrm{e}^{\mathrm{i}k_0\epsilon_{1s}z/\gamma_0} + \zeta_{2s}^{0}\mathrm{e}^{\mathrm{i}k_0\epsilon_{2s}z)/\gamma_0}\,, \\ \zeta_{1,2s}^{0} &= \mp\frac{2\epsilon_{2,1s} - \chi_{00}}{2(\epsilon_{2s} - \epsilon_{1s})}\,, \quad \zeta_{1,2s}^{g} = \mp\frac{\chi_{01}^{s}}{2(\epsilon_{2s} - \epsilon_{1s})}\,. \end{aligned} \tag{A.20}$$

(2b) The Laue case ($\beta > 0,\ \gamma_0 < 0,\ \gamma_g < 0$):

$$\begin{aligned}\Phi^{(s)}(z) &= H(-z) + D_{s0}^{(3*)}(z)H(z)H(L-z) + D_{s0}^{(3*)}(L)H(z-L)\,, \\ \Phi_g^{(s)}(z) &= \beta\left\{D_{sg}^{(3*)}(z)H(z)H(L-z) + D_{sg}^{(3*)}(L-z)H(z-L)\right\}\,, \\ D_{s0}^{(3)}(z) &= -\zeta_{1s}^{0}\mathrm{e}^{\mathrm{i}k_0\epsilon_{1s}z/|\gamma_0|} - \zeta_{2s}^{0}\mathrm{e}^{\mathrm{i}k_0\epsilon_{2s}z/|\gamma_0|}\,, \\ D_{sg}^{(3)}(z) &= \zeta_{1s}^{g}\mathrm{e}^{\mathrm{i}k_0\epsilon_{1s}z/|\gamma_0|} - \zeta_{2s}^{0}\mathrm{e}^{\mathrm{i}k_0\epsilon_{2s}z)/|\gamma_0|}\,. \end{aligned} \tag{A.21}$$

## A.3 Accurate Calculation of PXR with Multiple Scattering of Electrons

For the description of PXR fine structure and high-resolution PXR, a more accurate calculation than that in Sect. 2.3 of the multiple scattering of charged particles is necessary. Equation (2.16) has to be averaged over all the particle trajectories in the crystal [4, 5]:

$$\begin{aligned}W_{\boldsymbol{n}\omega}^{(s)} &= \frac{q^2\omega^2}{4\pi^2c^3}\int_{-\infty}^{\infty}\mathrm{d}t\int_{-\infty}^{\infty}\mathrm{d}t'\; w_1(\boldsymbol{r},\boldsymbol{v},t)w_2(\boldsymbol{r},\boldsymbol{v},t|\boldsymbol{r}',\boldsymbol{v}',t') \\ &\quad \mathrm{e}^{\mathrm{i}\omega(t-t')}\left(\boldsymbol{v}\boldsymbol{E}_{\boldsymbol{k}s}^{(-)}(\boldsymbol{r},\omega)\right)^{*}\left(\boldsymbol{v}'\boldsymbol{E}_{\boldsymbol{k}s}^{(-)}(\boldsymbol{r}',\omega)\right)\,, \end{aligned} \tag{A.22}$$

where $w_1(\boldsymbol{r},\boldsymbol{v},t)$ is the particle distribution function at the time $t$, $w_2(\boldsymbol{r},\boldsymbol{v},t|\boldsymbol{r}',\boldsymbol{v}',t')$ is the probability density to find a particle at the time $t'$ at the position $\boldsymbol{r}',\boldsymbol{v}'$, if it was at the position $\boldsymbol{r},\boldsymbol{v}$ at the time $t$.

The periodic crystal structure influences the beam distribution function in a small phase volume near the boundaries of the Brillouin zones. This case is essential if the primary beam velocity $\boldsymbol{v}_0$ is parallel to the crystallographic axes (planes) and particles are trapped into the channelling mode. In other cases, the kinetic equation for a homogeneous medium can be used for averaging the distribution function over the trajectories. The energy $E$ and $\vartheta$ of velocity deviation are more convenient variables to be used in $w_1, w_2$ instead of the velocity $\boldsymbol{v}$. In the case of relativistic particles $\vartheta \ll 1$, the kinetic equation is [4]

$$\frac{\partial w}{\partial t} + \boldsymbol{v}\frac{\partial w}{\partial \boldsymbol{r}} = q(E)\Delta_\vartheta w + \hat{K}(E)w \; ;$$
$$q(E) = \frac{c}{L_R}\frac{E_\mathrm{s}^2}{4E^2}, \qquad \Delta_\vartheta = \frac{\partial^2}{\partial \vartheta_x^2} + \frac{\partial^2}{\partial \vartheta_y^2} \, ,$$
$$\hat{K}(E)w = \int_0^\infty \frac{u^2 + E^2 - 2uE/3}{(u-E)}\left[\frac{w(\vartheta, u, t)}{u^2}H(u-E) + \frac{w(\vartheta, E, t)}{E^2}H(E-u)\right] . \tag{A.23}$$

Equation (A.23) uses the Bete–Gaitler [2] formula for bremsstrahlung, which is dominant in particle energy losses; the parameter $E_\mathrm{s}$ and $L_R$ from (1.26) are the characteristic energy and the radiation length of the multiple scattering of electron beam on the shielded potential of crystal atoms, respectively. The functions $w_1$ and $w_2$ are the solutions of (A.23) under different initial conditions:

$$w_1(t=0) = \delta(\boldsymbol{r} - \boldsymbol{r}_0)\delta(\vartheta_x)\delta(\vartheta_y)\delta(E - E_0) \, ,$$
$$w_2(t=t') = \delta(\boldsymbol{r} - \boldsymbol{r}')\delta(\vartheta_x - \vartheta'_x)\delta(\vartheta_y - \vartheta'_y)\delta(E - E') \, , \tag{A.24}$$

where $\boldsymbol{r}_0$ and $E_0$ are the initial position and the beam energy at the time $t = 0$, respectively.

To analyse the radiation spectrum in a crystal of thickness $L$, the expressions for wave fields from Appendix A.2 have to be used. The integration area is divided into three parts: $(-\infty, 0)$, $(0, t_0)$, $(t_0, \infty)$, where $t_0 = L/(\boldsymbol{v}_0\boldsymbol{N})$ corresponds to the time of the escape of the particle from the crystal and the fluctuations of the time taken by the particle to pass through the crystal due to multiple scattering are neglected; $\boldsymbol{N}$ is a normal to the crystal surface. Thus, there are nine different contributions to the total radiation intensity, each having a certain physical interpretation and depending on the experimental geometry. The detailed analysis of (A.22) is given in [5], and an example of the Laue geometry is presented below. For relativistic particles, the energy losses for radiation are comparatively low, whereas multiple scattering is essential. Then the operator $\hat{K}(E)$ in (A.23) can be dropped and $q(E) = q(E_0) \equiv q_0$ is constant, and the solution of (A.23) for $w_1(\boldsymbol{r}, \vartheta, t), w_2(\boldsymbol{r} - \boldsymbol{r}', \vartheta, \vartheta', t - t')$ with the initial conditions (A.24) are derived in [13, 14]. The contribution to the intensity (A.22) due to the particle trajectory in the crystal is [5]

$$W_{\boldsymbol{n}\omega}^{(s)} = \frac{q^2\omega^2}{2\pi^2 c^3} C_\mathrm{s}^2 \left|\frac{\chi_g}{2(\epsilon_{2s} - \epsilon_{1s})}\right|^2 \mathrm{Re} \int_0^{t_0} \mathrm{d}t \int_0^{t_0 - t} \mathrm{d}\tau$$
$$\times \; \left[(\boldsymbol{e}_{gs}\boldsymbol{v}_0)^2 Q_1(1+\Delta) + 2q_0 t Q_2\right]\left[\sum_{\mu=1,2} F_{\mu s} - \Phi_\mathrm{s}\right] \, , \tag{A.25}$$

where the polarization factor $C_\mathrm{s}$ and the diffraction parameters follow from A.2, and the multiple scattering is determined by

$$Q_1 = \frac{1}{\cosh u(1+\eta\tau\tanh u)} \qquad \exp\left[\frac{\mathrm{i}\omega\tau\theta^2}{2} - \frac{\eta\theta^2\tanh u}{4q(1+\eta t\tanh u)}\right],$$

$$Q_2 = \frac{Q_1}{\cosh u(1+\eta\tau\tanh u)}, \qquad \Delta = \frac{1-\cosh u(1+\eta\tau\tanh u)^2}{\cosh u(1+\eta\tau\tanh u)^2},$$

$$F_{\mu s} = \exp\left[-\frac{\mathrm{i}c\tau}{L_{\mu s}} - \frac{t_0 - t - \tau/2}{L_{\mu s}^{(a)}}\right],$$

$$\Phi_s = \exp\left[-\frac{\mathrm{i}c\tau}{L_{2s}} + \frac{\tau}{L_{2s}^{(a)}} - \mathrm{i}\omega(\epsilon_{2s}^* - \epsilon_{1s})(t_0 - t)\right]$$
$$+ \exp\left[-\frac{\mathrm{i}c\tau}{L_{2s}} - \frac{\tau}{L_{2s}^{(a)}} + \mathrm{i}\omega(\epsilon_{2s} - \epsilon_{1s}^*)t\right],$$

$$u = \eta\tau, \quad \eta = \sqrt{2\mathrm{i}\omega q_0},$$

$$L_{\mu s} = \frac{2c}{\omega[\gamma^{-2} + \theta^2 + 2(1 - \mathrm{Re}\ \epsilon_{\mu s})]}, \qquad L_{\mu s}^{(a)} = \frac{2c}{\omega\ \mathrm{Im}\ \epsilon_{\mu s}}. \tag{A.26}$$

The functions $Q_1$, $Q_2$ and $\Delta$ depend on the coherent length of bremsstrahlung [14]:

$$L_{\mathrm{BS}} = \frac{c}{\sqrt{2\omega q_0}},$$

and their influence on the spectral–angular characteristics of radiation is determined by the ratio of $L_{\mathrm{BS}}$ and PXR coherent length $L_{\mu s}$.

For high-energy electrons ($L_{\mathrm{BS}} \sim \gamma \to \infty$) or for heavy charged particles ($q_0 \to 0$), the functions $Q_{1,2} \to 1$, $\Delta \to 0$. Then the main contribution to the intensity is given by the first term in (A.25), which has a minimum $\sim \theta^2$ of the photon radiation in the diffraction direction $\boldsymbol{k}_{\mathrm{B}}$. The term in (A.25) proportional to $2q_0tQ_2$ corresponds to bremsstrahlung, which has a maximum in the direction $\boldsymbol{k}_{\mathrm{B}}$. In the case of thin crystals, these facts fit well the results of Sect. 2.3. In general, the influence of multiple scattering results in cumbersome expressions and has been investigated in [5]. Here we emphasize only the expression which is useful for fitting of the HRPXR experimental data (see Sect. 2.3). The formula used in (3.26) for the photon radiation angle,

$$\theta_{\mathrm{ph}}^2 = \gamma^{-2} + \theta_{\mathrm{sc}}^2 + \theta_{\mathrm{M}}^2,$$

has to be substituted for higher accuracy by

$$\theta_{\mathrm{ph}}^2 = \gamma^{-2} + \zeta\theta_{\mathrm{sc}}^2 + \theta_{\mathrm{M}}^2,$$

and the dimensionless parameter $\zeta$ is varied in the region $1.2 < \zeta < 3$ for different crystal thicknesses [5].

## References

1. A.M. Afanas'ev, Yu.M. Kagan: Zh. Eksp. Teor. Fiz. **48**, 327 (1965)

2. A.I. Akhiezer, V.B. Berestetzkii: *Quantum Electrodynamics* (Nauka, Moscow 1969) pp 463–467
3. A. Authier: *Dynamical Theory of X-ray Diffraction* (Oxford University Press, New York 2001) pp 3–26
4. V.G. Baryshevsky: *Channelling, Radiation and Reactions at Crystals under High Energy* (Belarussian State University, Minsk 1982)
5. V.G. Baryshevsky, A.O. Grubich, Le T'en Hai: Zh. Eksp. Theor. Fiz. **94**, 51 (1988)
6. B. Batterman, H. Cole: Rev. Mod. Phys. **36**, 681 (1964)
7. I.D. Feranchuk, L.I. Gurskii, L.I. Komarov, O.M. Lugovskaya, F. Burgäzy, A.P. Ulyanenkov: Acta Crystallogr. A **58**, 370 (2002)
8. *International Tables for Crystallography*, vol C (Kluwer, Dodrecht 1992)
9. R.W. James: *The Optical Principles of the Diffraction of X-rays* (G. Bell and Sons, London 1950)
10. L.D. Landau, E.M. Lifshitz: *Electrodynamics of Continuous Media* (Nauka, Moscow 1982)
11. T.C. Marshall: *Free-Electron Lasers* (Macmillan, New York, London 1985)
12. P. Morse, H. Feshbach: *Methods of Theoretical Physics* (McGraw-Hill, New York 1953) p 735
13. V.E. Pafomov: Zh. Tekhn. Fiz **33**, 557 (1963) (Sov. Phys. Tech. Phys. **8**, 412 (1963)); Proc. Lebedev Inst. Phys. **44**, 25 (1971)
14. M.L. Ter-Mikaelian: *High Energy Electromagnetic Processes in Condensed Media* (in Russian: AN ArmSSR, Yerevan 1969) (in English: Wiley, New York 1972)

# Index

# Springer Tracts in Modern Physics

185 **Electronic Defect States in Alkali Halides**
Effects of Interaction with Molecular Ions
By V. Dierolf 2003. 80 figs., XII, 196 pages

186 **Electron-Beam Interactions with Solids**
Application of the Monte Carlo Method to Electron Scattering Problems
By M. Dapor 2003. 27 figs., X, 110 pages

187 **High-Field Transport in Semiconductor Superlattices**
By K. Leo 2003. 164 figs.,XIV, 240 pages

188 **Transverse Pattern Formation in Photorefractive Optics**
By C. Denz, M. Schwab, and C. Weilnau 2003. 143 figs., XVIII, 331 pages

189 **Spatio-Temporal Dynamics and Quantum Fluctuations in Semiconductor Lasers**
By O. Hess, E. Gehrig 2003. 91 figs., XIV, 232 pages

190 **Neutrino Mass**
Edited by G. Altarelli, K. Winter 2003. 118 figs., XII, 248 pages

191 **Spin-orbit Coupling Effects in Two-dimensional Electron and Hole Systems**
By R. Winkler 2003. 66 figs., XII, 224 pages

192 **Electronic Quantum Transport in Mesoscopic Semiconductor Structures**
By T. Ihn 2003. 90 figs., XII, 280 pages

193 **Spinning Particles – Semiclassics and Spectral Statistics**
By S. Keppeler 2003. 15 figs., X, 190 pages

194 **Light Emitting Silicon for Microphotonics**
By S. Ossicini, L. Pavesi, and F. Priolo 2003. 206 figs., XII, 284 pages

195 **Uncovering *CP* Violation**
Experimental Clarification in the Neutral K Meson and B Meson Systems
By K. Kleinknecht 2003. 67 figs., XII, 144 pages

196 **Ising-type Antiferromagnets**
Model Systems in Statistical Physics and in the Magnetism of Exchange Bias
By C. Binek 2003. 52 figs., X, 120 pages

197 **Electroweak Processes in External Electromagnetic Fields**
By A. Kuznetsov and N. Mikheev 2003. 24 figs., XII, 136 pages

198 **Electroweak Symmetry Breaking**
The Bottom-Up Approach
By W. Kilian 2003. 25 figs., X, 128 pages

199 **X-Ray Diffuse Scattering from Self-Organized Mesoscopic Semiconductor Structures**
By M. Schmidbauer 2003. 102 figs., X, 204 pages

200 **Compton Scattering**
Investigating the Structure of the Nucleon with Real Photons
By F. Wissmann 2003. 68 figs., VIII, 142 pages

201 **Heavy Quark Effective Theory**
By A. Grozin 2004. 72 figs., X, 213 pages

202 **Theory of Unconventional Superconductors**
By D. Manske 2004. 84 figs., XII, 228 pages

203 **Effective Field Theories in Flavour Physics**
By T. Mannel 2004. 29 figs., VIII, 175 pages

204 **Stopping of Heavy Ions**
By P. Sigmund 2004. 43 figs., XIV, 157 pages

205 **Three-Dimensional X-Ray Diffraction Microscopy**
Mapping Polycrystals and Their Dynamics
By H. Poulsen 2004. 49 figs., XI, 154 pages

206 **Ultrathin Metal Films**
Magnetic and Structural Properties
By M. Wuttig and X. Liu 2004. 234 figs., XII, 375 pages

207 **Dynamics of Spatio-Temporal Cellular Structures**
Henri Benard Centenary Review
Edited by I. Mutabazi, J.E. Wesfreid, and E. Guyon 2005. approx. 50 figs., 150 pages

208 **Nuclear Condensed Matter Physics with Synchrotron Radiation**
Basic Principles, Methodology and Applications
By R. Röhlsberger 2004. 152 figs., XVI, 318 pages

209 **Infrared Ellipsometry on Semiconductor Layer Structures**
Phonons, Plasmons, and Polaritons
By M. Schubert 2004. 77 figs., XI, 193 pages

210 **Cosmology**
By D.-E. Liebscher 2005. Approx. 100 figs., 300 pages

211 **Evaluating Feynman Integrals**
By V.A. Smirnov 2004. 48 figs., IX, 247 pages

213 **Parametric X-ray Radiation in Crystals**
By V.G. Baryshevsky, I.D. Feranchuk, and A.P. Ulyanenkov 2006. 63 figs., IX, 172 pages

Printing: Krips bv, Meppel
Binding: Stürtz, Würzburg